LIGHT SOURCES
Technologies and Applications

LIGHT SOURCES

Technologies and Applications

Spiros Kitsinelis

CRC Press
Taylor & Francis Group
Boca Raton London New York

CRC Press is an imprint of the
Taylor & Francis Group, an **informa** business

A TAYLOR & FRANCIS BOOK

Taylor & Francis
6000 Broken Sound Parkway NW, Suite 300
Boca Raton, FL 33487-2742

© 2011 by Taylor and Francis Group, LLC
Taylor & Francis is an Informa business

Printed in the United States of America on acid-free paper
10 9 8 7 6 5 4 3 2 1

International Standard Book Number: 978-1-4398-2079-7 (Hardback)

Library of Congress Cataloging-in-Publication Data

Kitsinelis, Spyridon.
    Light sources : technologies and applications / Spiros Kitsinelis.
        p. cm.
    Summary: "This book provides a description of modern light sources, covering their uses in a range of sciences and industries as well as the relevant scientific and engineering principles. The authors discuss incandescent and discharge sources along with more modern lasers and solid state light sources. They also explore various applications across a range of areas, from general lighting to specialized uses that require specific emissions, such as in sterilization, health science, medicine, food processing, and spectroscopy. The book is essential for those dealing with light sources and their applications in spectroscopy, optics, and industry"-- Provided by publisher.
    Includes bibliographical references and index.
    ISBN 978-1-4398-2079-7 (hardback)
    1. Lasers. 2. Light sources. 3. Optics. I. Title.

TA1520.K58 2010
621.32--dc22   2010037000

Visit the Taylor & Francis Web site at
http://www.taylorandfrancis.com

and the CRC Press Web site at
http://www.crcpress.com

Table of Contents

Preface

This book is an overview of all known families of light sources and the scientific principles they are based on. The book offers an introduction to the main technologies of existing products as well as a deeper analysis of subjects such as the nature of light, the processes of creating it, and the appropriate use of lamps in different applications in order to satisfy both the professional and the needs of everyday life. This book will be essential to a range of scientists and students dealing with spectroscopy and light source subjects, as well as other professionals such as engineers, architects, decorators, artists, electricians, lighting designers, and many others in the lighting industry.

Author's Bio

Dr. Spiros Kitsinelis is a researcher focusing on the development of novel and energy efficient light sources. He gained his master's and Ph.D. degrees in chemistry from the University of Sheffield in England for his research and development of pulse-operated low-pressure plasma light sources in the High Temperature Science Laboratories. He continued his research as a post-doctoral fellow at Ehime University in Japan in the department of electrical and electronic engineering. He held the position of Project Leader, at Philips Lighting Central Development Lighting, in the Netherlands, and he continued his research and development of the next generation of plasma light sources for the physics department of the National Technical University of Athens, Greece. After a break from research when he served as a chemical engineer for the armed forces he returned to civilian life and his research activities. Among other posts he also acted as the National Contact Point for Energy at the National Documentation Centre of the National Research Foundation of Greece. He is the author of a number of scientific publications, has attended many international conferences, and is the cocreator of a number of patents in the field of light sources, as well as the editor of the electronic periodical *Science and Technology of Light Sources (SATeLightS)*.

CHAPTER 1

Introduction

HUMANITY HAS TO DEAL with two main issues regarding energy. The first is the availability of nonsustainable energy sources and whether the global demand for energy can be met. This is due to their depletion in certain parts of the world or due to geopolitical factors and, in any case, their impact on the global economy is substantial. The second issue is environmental changes and their impact on our lives. The burning of fossil fuels as the most common energy generation mechanism results in the formation and emission of carbon dioxide as a by-product, which is one of the gases responsible for the greenhouse effect.

Considering that humans are using about a fifth of the world's generated electric energy for lighting applications [1,2], it is easy to appreciate the importance of light source technology both from economic and environmental standpoints. Light sources and lighting not only represent an economic market of billions of euros, but the consumption of energy for lighting is responsible for the generation of millions of tons of CO_2 gas annually.

Furthermore, light is vital for life, and light sources play an indispensable role in our daily life. The quality of life and aspects such as health and urban security, which are related to traffic safety and city crime, depend on light and on its quality. And light sources are not limited to general lighting but are also applied in a range of other applications that require emissions in the ultraviolet and infrared such as sterilization, health science, aesthetic medicine, art conservation, food processing, and sterilization of hospitals or water to name a few.

Nobody can imagine a world without light, where darkness dominates and one's mind conjures up both real and fantastic threats. We would

1

surely be talking of a different world. Light has been bound up in our consciousness with our very existence, and our efforts to create it using our own technical means, as well as to understand its nature, began thousands of years ago when humans first used fire. Over time, the burning of wood was replaced by the burning of oils and, later, in the 18th century, by the burning of gases. The harnessing of electricity brought about a revolution not only in the way we live our lives but also in the way we light up our lives. From the middle of the 19th century, various methods of generating light by employing electricity were devised, and these technologies kept evolving through the decades.

This book focuses on the three main technologies of creating light via electricity and the various families of light sources that were born out of them. The first technology, incandescence of a filament, was due to the efforts of people such as Heinrich Gobel in the middle of the 19th century and of Joseph Swan and Thomas Edison a few years later. The second technology, electrical discharges through gases, took off in the beginning of the 19th century thanks to Humphry Davy. The third technology, diodes, the result of developments in the semiconductor field, were born much later in the middle of the 20th century, once again revolutionizing the field of lighting. Before we start discussing the principles of these three technologies and the families of light sources that they gave rise to, we shall take a look at the nature of light and how we perceive it and study it. When Isaac Newton in the middle of the 17th century analyzed white light into its constituent colors, explaining the formation of rainbows, he did not take away the magic of this phenomenon, but opened the door to another magical world that had to do with the nature of light.

And although since the times of the ancient Greek philosophers, questions on the way the human eye functions and the nature of light have tantalized scientists, today, after centuries of experiments and scientific disputes, certain ideas and theories have been proved and become universally accepted.

This book contains information on representative products of each family of light sources, but it is not an exhaustive list of all available commercial products. There would be no point in trying to keep up with all special characteristics of new products that change rapidly, and this kind of updating would be better served by the catalogs of all major companies. This book is an overview of all main technologies and important families of light sources that have dominated not only the markets but our lives since the end of the 19th century.

The photographic material of this book comes from three main sources:

- The author's own collection
- http://www.lamptech.co.uk/ by J.D. Hooker
- http://commons.wikimedia.org

The reader will find references on specific research results at the end of chapters, but the general information presented throughout this book is based on the following two important sources:

- *Lamps and Lighting* by J.R. Coaton and A.M. Marsden ISBN 0-340-64618-7.
- *Electric Discharge Lamps* by John Waymouth ISBN 0-262-23048-8.

REFERENCES

1. DG—TREN, EU energy and transport in figures, Statistical book 2007/2008.
2. IEA (2006) Light's labor's lost.

CHAPTER 2

Basic Principles of Light and Vision

2.1 NATURE AND PROPERTIES OF LIGHT

After the mid-17th century, scientists were divided into two sides. One side, including Isaac Newton, believed in the corpuscular theory of light and talked of effects such as reflection and shadows to support their views, while the other side, led by Christian Huygens, believed in the wave properties of light as shown by phenomena such as diffraction. In the early 20th century, the conflict was resolved when scientists, particularly Albert Einstein and Louis de Broglie, gave a new picture of quantum physics by showing the duality of matter and energy at the atomic level. A series of experiments showed that photons act like particles, as demonstrated by the photoelectric effect and Compton scattering—but also as waves. This property of theirs remains today one of the most interesting and bizarre aspects of the natural world.

Regarding the wave property of photons, we can talk about electromagnetic waves, which are characterized by intensity, wavelength, and polarization. Electromagnetic waves are transverse waves where the oscillation is perpendicular to the direction of travel, as we know from James Maxwell's equations and the experiments of Heinrich Hertz. Accordingly, an electric field changing in time creates a magnetic field, and vice versa. These two fields oscillate perpendicular to each other and perpendicular to the direction of motion of the wave (Figure 2.1).

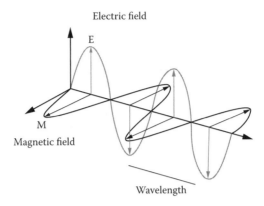

FIGURE 2.1   Electromagnetic wave propagation.

Regardless of wavelength, waves travel at the same speed in vacuum (299.792458 km/s), and the electromagnetic spectrum extends from radio waves with wavelengths of up to several kilometers to cosmic rays with wavelengths of a few angstroms (10^{-10} m). The relationship between energy and wavelength or frequency is given by the formula

$$E = h \cdot v = (h \cdot c)/\lambda$$

where

E = energy (Joule)
v = frequency (Hz)
h = Planck constant (6.626×10^{-34} J·s)
c = speed of light in vacuum (2.998×10^{8} m/s)
λ = wavelength (m)

Visible light is just a small part of the electromagnetic spectrum to which the human eye is sensitive and includes waves with lengths from around 380 nm (10^{-9} m) to about 780 nm. On the lower-energy side, the entire spectrum starts with radio waves, which we use to transfer images and sounds like radio and television; continues to microwaves, used in devices such as radar and the known oven for domestic use; and further down, one finds infrared waves, which we perceive as heat. On the higher-energy side of the visible spectrum with shorter wavelengths, we find ultraviolet radiation, which is divided into three parts depending on the wavelength.

The first part is the one that causes skin tanning and helps us absorb some vitamins. The second part is dangerous and causes cancer, while the third part of the shorter wavelengths is extremely dangerous to human health and is blocked by the atmosphere. Then, we find x-rays used in medicine for bone outlining; gamma rays; and, finally, cosmic rays. The different regions of the electromagnetic spectrum are presented in the appendix.

During propagation, light can be reflected, and the kind of reflection depends on the reflecting material. In specular reflection, the angle of reflection equals the angle of incidence. Materials such as aluminum and silver are mainly used for manufacturing mirrors, and they show the same reflectivity to all the wavelengths of the visible spectrum, while for diffuse (Lambertian) reflection, such as the one we find in an integrating sphere (Ullbricht spheres), barium sulfate ($BaSO_4$) is used.

Its wave nature gives light some other interesting properties, such as refraction and interference. It is these properties that several spectrometers make use of in order to analyze the visible light and radiation in general.

The principle of refraction is a change of direction following a change in the speed of the waves, and that happens when a wave passes from one medium to another with different optical density, or as we call it, a different refractive index. Figure 2.2 depicts this change of direction when the medium changes.

The angle of incidence and the angle of refraction are related to the refractive indices of the media by Snell's law:

$$n_1\sin\theta_1 = n_2\sin\theta_2$$

where n is the refractive index of each medium, and $\sin\theta$ is the sine of each angle (for example, air has a refractive index of 1.0003; for water it is 1.33 and for glass it is 1.5–1.9, depending on the type of glass).

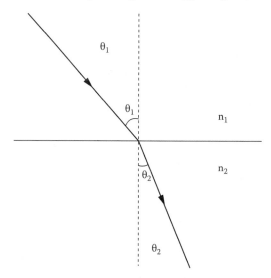

FIGURE 2.2   Wave refraction.

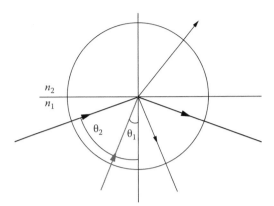

FIGURE 2.3   Total internal reflection.

If the angle of incidence exceeds a specific value, then the wave is totally reflected without refraction taking place. This effect, which can be observed only for waves traveling from a medium of higher refractive index to one of lower refractive index (glass to air, not vice versa), is known as *total internal reflection*, and it is the principle on which optical fibers work (Figure 2.3).

Refraction underlies the optical properties of lenses. The Fresnel design allows the construction of large lenses by reducing the volume and weight that would be required for the construction of conventional lenses (Figure 2.4).

Refraction is also responsible for the effect of the rainbow and the splitting of white light into component colors through a prism. The different frequencies of different colors travel at different speeds when they enter a new medium, leading to different directions for each wave of different color. This is also called chromatic aberration and can be seen in lenses of inferior quality.

This means that the wavelength affects the refractive index of the medium. The color of the sky at sunset, and changes in the color of the moon are the result of refraction, where the earth's atmosphere acts as the second optical medium.

Figure 2.5 shows the refraction of white light through a prism. Each color is refracted to different degrees and, therefore, the waves bend to different degrees, resulting in the splitting of the different colors.

The other interesting property of light waves is diffraction: When waves pass obstacles or find an opening of similar dimensions to their

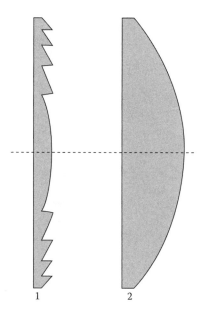

FIGURE 2.4 (1) Fresnel lens and (2) conventional lens.

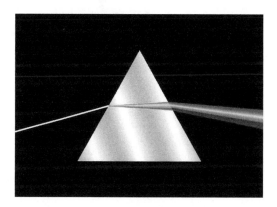

FIGURE 2.5 **(See color insert following page 20.)** Prism diffraction: analyzing white light to its constituent colors.

wavelengths, they spread, and interference occurs as shown in Figure 2.6. The interference results in new waves of intensity that are equal to the sum of the intensities of the initial waves at each point. This could mean an increase, or the zeroing, of intensity at some point of the new wave. Figure 2.7 shows schematically this summation of wave amplitudes.

Figures 2.8 and 2.9 show the effect of diffraction as seen in everyday life and how it is used for the development of scientific instruments.

FIGURE 2.6 Wave interference.

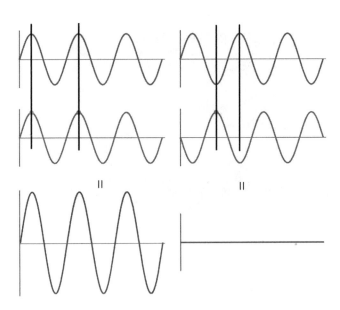

FIGURE 2.7 The wave interference can be (left) constructive or (right) destructive, depending on the phase difference of the waves.

FIGURE 2.8 **(See color insert following page 20.)** The diffraction of light when incident on surfaces such as a compact disk, which consists of several grooves, leading to interference and the formation of colors.

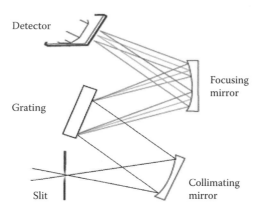

FIGURE 2.9 Spectrometers use the diffraction effect in order to analyze light into its component wavelengths.

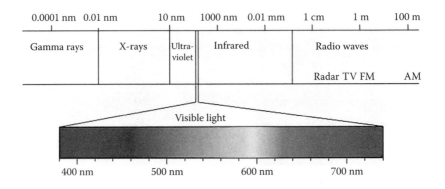

FIGURE 2.10 **(See color insert following page 20.)** The electromagnetic spectrum.

Visible light is part of the electromagnetic spectrum and consists of a series of waves with lengths from 380 to 780 nm that humans perceive as color (Figure 2.10).

2.2 VISION

The human eye operates almost like a camera. The iris controls the amount of light that passes through, and the lens focuses and projects the image upside down at the back end of the eye. A series of sensors are activated, and the image is transferred with the help of neurotransmitters to the brain for processing. Problems of myopia or presbyopia occur when the lens is unable to focus on the right spot or when the size of the eye is such that correct focusing is again not possible (Figure 2.11).

Specifically, the way we see an object is through the following steps.

Light passes through the cornea, the pupil, the iris, the lens, the vitreous humor, the retina, the optic nerve, the optic pathway, the occipital cortex, and, finally, reaches the brain, where processing takes place. Figure 2.12 depicts the anatomy of the human eye.

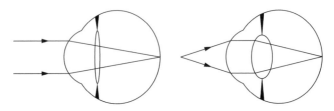

FIGURE 2.11 Vision problems occur when the light from a source is not focused at the correct spot of the eye.

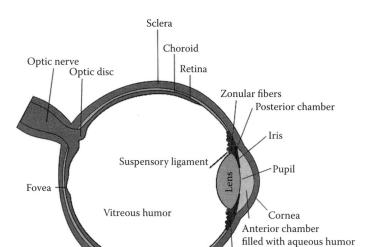

FIGURE 2.12   Anatomy of the human eye.

The human eye has two different kinds of photoreceptor cells, cones and rods. Cones are responsible for color perception (three different groups of cones for three different parts of the visible spectrum), while rods are more sensitive to light intensity so they are mostly activated during the night or in low-light conditions. Rods are faster and more sensitive than cones. Color is further broken down into "hue" and "saturation." The eye, as a photoreceptor, is sensitive to only a narrow band of the electromagnetic spectrum, this band corresponding to "visible light" that stretches from 360 (violet) to 780 nm (red).

Figure 2.13 shows the sensitivity curves representing each of the three groups of color-sensing cells in the eyes (the cones). Cone cells, or cones, are photoreceptor cells in the retina of the eye that function best in relatively bright light. The cone cells gradually become sparser toward the periphery of the retina. Cones are less sensitive to light than the rod cells in the retina (which support vision at low-light levels), but allow the perception of color. Figure 2.14 shows the overall sensitivity of the human eye in different light level conditions.

Thus, the lumen should be considered a "weighted" power unit, taking into account the human eye's sensibility. Photopic vision is the only one of interest to lighting design because the eye is "light-adapted" rather than "dark-adapted" under most illumination levels produced by artificial light sources. However, an intermediate situation called "mesopic" vision

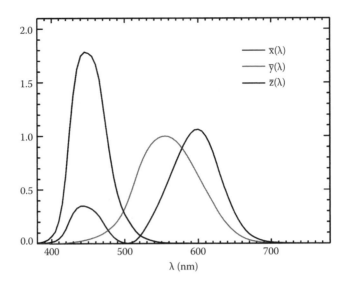

FIGURE 2.13   Human eye sensitivity curves.

corresponding to dusk conditions might prove to be a determinant factor for achieving energy savings, especially in street lighting systems.

Mesopic vision describes the transition region from rod vision (scotopic) to cone vision (photopic), where signals from both rods and cones contribute to the visual response. Mesopic vision covers approximately

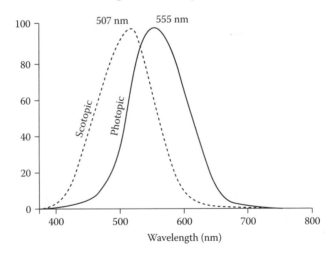

FIGURE 2.14   The figure shows the eye sensitivity curve of the eye, which peaks at 555 nm. This peak shifts toward blue when the light is at low levels (rods are more sensitive to blue light).

4 log units and encompasses a range of light levels often found in occupational environments. The performance of light sources is often compared by reference to their efficacy, that is, output measured in lumens per watt of input power. However, the definition of the lumen is derived from the spectral luminous efficiency function for photopic vision only: The eye's sensitivity is described by standards published by the Commission Internationale de l'Eclairage (CIE) for photopic and scotopic conditions. Under photopic conditions, the sensitivity of the human eye peaks at 555 nm. As the luminance decreases, the peak of the sensitivity shifts toward lower wavelengths. The peak sensitivity under scotopic conditions is at 507 nm. These data are known as the spectral luminous efficiency functions or the $V(\lambda)$ curves. There is not an equivalent standard for the mesopic region yet, and there will be developments in this area soon.

At the maximum sensitivity of the photopic curve, which is at 555 nm, a radiant watt emitted by a source corresponds to 683 lumens (lm).

A wide emission spectrum is called continuous, while atomic sharp emissions or emissions in narrow ranges give us a line spectrum.

White daylight is composed of all colors, or wavelengths, as shown in the following continuous spectrum of Figure 2.15.

Nevertheless, it is possible to create white light without the need for all wavelengths, but with only one wavelength from each region of the spectrum that stimulates each of the three groups of sensory organs of the eye, namely, the cones. More specifically, we could create the impression of white light with a combination of red, green, and blue radiation, which are the colors we call primary, as shown in Figure 2.16. However, media that use reflected light and its partial absorption in order to create colors are using the abstractive method of mixing colors, as shown in Figure 2.17.

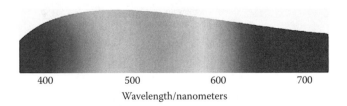

400   500   600   700

Wavelength/nanometers

FIGURE 2.15 **(See color insert following page 20.)** The color analysis/spectrum of daylight.

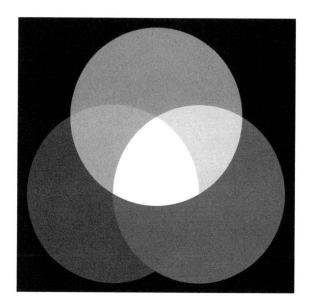

FIGURE 2.16 **(See color insert following page 20.)** The combination of the three primary colors gives white light. Any color can be created with the appropriate combination of primary colors.

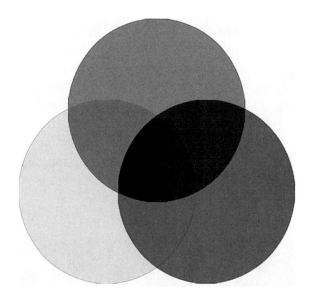

FIGURE 2.17 **(See color insert following page 20.)** Media that use reflected light and its partial absorption in order to create colors are using the abstractive method of mixing colors.

2.3 LIGHT SOURCE CHARACTERISTICS

The color rendering index (CRI) is a measure of how well the light source reproduces the colors of any object in comparison to a reference source. The CRI of a lamp is obtained by measuring the fraction of light reflected from each of a number of surfaces of specific color samples covering the visible spectrum. We arbitrarily attributed a maximum CRI of 100 to the light source that most closely reproduces colors as illuminated by standard white light. Some representative values are given in Figure 2.18.

An issue to be addressed is the color rendering of the objects, which is not the same when one uses daylight compared to white light made up from the three primary colors. A CRI has therefore been defined on a scale of 0 for monochromatic sources to 100 for sources that emit a continuous visible spectrum. The less the visible parts of the spectrum covered by the radiation source, the smaller the CRI.

Differences in the CRI between two white light sources lead to phenomena such as metamerism. A fabric, for example, whose color is seen correctly under the light of a source emitting a continuous spectrum, can change color when illuminated by another source of white light that does not emit one of the colors of the fabric dye.

Apart from the percentage of the visible spectrum that the emitted radiation covers and which provides the CRI of the light source, what is also of importance is the intensity of the radiation in each color region. For example, a source that is emitting white light, but with larger percentages of radiation in the red part of the spectrum than the blue, gives a feeling

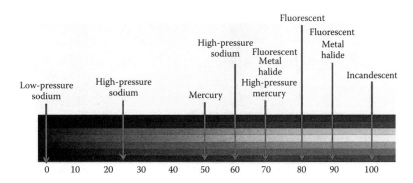

FIGURE 2.18 The color rendering index increases as more of the visible spectrum covered by emissions.

of warm white, and mostly blue radiation creates a cool white light. This property is called the *color temperature* and is directly linked to the emission of blackbody radiation.

A *blackbody* is defined as a body that has the potential to radiate at all wavelengths, with the percentage of radiation at shorter wavelengths increasing as the temperature of the body increases. The body initially emits infrared radiation and, as the temperature rises, there is emission at wavelengths that the human eye detects. The body initially emits red light, goes through orange, yellow, and blue, and finally reaches appears white since the emission has covered most of the visible spectrum (Figure 2.19). The relationship between body temperature and the emitted radiation is given by the following formulas.

According to Wien:

$$\lambda_{max} \,(nm) = 2.8978 \cdot 10^6 / T$$

Where the temperature T is in kelvin and the wavelength in nanometers, while the total radiated energy is given according to Stefan–Boltzmann's law.

$$E = \sigma \cdot T^4$$

Where E is measured in W/m², T in kelvin, and σ is the Stefan–Boltzmann constant.

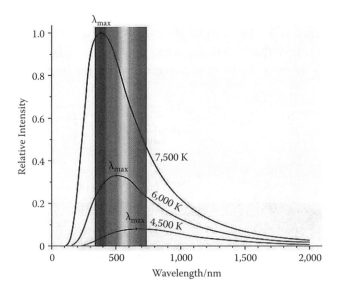

FIGURE 2.19 The maximum of a blackbody radiation shifts to smaller wavelengths as the temperature rises.

1800 K	4000 K	5500 K	8000 K	12000 K	16000 K

FIGURE 2.20   Scale of color temperature.

Thus, the color temperature of a light source is the temperature in kelvin units in which a heated blackbody acquires the same color tone as the source. Sources are usually called warm when the color temperatures are below 3300 K and cold when at color temperatures above 5000 K (Figure 2.20).

Different lamps are characterized by different color temperatures and CRIs as shown in Table 2.1 and Figures 2.21, 2.22, and 2.23.

TABLE 2.1   Indicative Values of Color Rendering Indices and Color Temperatures for Various Lamps and Light Source Technologies

Light Source	CCT (K)	CRI
Mercury	6410	17
Sodium high pressure	2100	25–85
Warm white fluorescent	2940	55–99
Cool white fluorescent	4080	55–99
Metal halide	5400	70–90
Incandescent	3200	100

FIGURE 2.21   **(See color insert following page 20.)** Different lamps are characterized by different color temperatures and color rendering indices.

FIGURE 2.22 **(See color insert following page 20.)** Different lamps are characterized by different color temperatures and color rendering indices.

Since the human eye is using three groups of photoreceptors for three different regions of the visible spectrum, a diagram construction describing all colors would be three dimensional. After studies and experiments by W. David Wright (Wright 1928) and John Guild (Guild 1931), the first chromaticity diagram was created. For convenience, the international committee on illumination (CIE) agreed on two color dimensions (color coordinates) and one dimension of intensity. In this way, a two-dimensional chromaticity diagram (Figure 2.24) was defined at the maximum intensity of the original diagram. This diagram allows us to see the color that the human eye will perceive from a given light source with a specific emission spectrum.

For some light sources that do not emit continuous radiation throughout the visible part of the spectrum as a blackbody does, a *correlated color temperature (CCT)* is assigned, which is defined as the point on the blackbody locus that is nearest to the chromaticity coordinates of the source (Figure 2.25).

FIGURE 2.23 **(See color insert following page 20.)** Different lamps are characterized by different color temperatures and color rendering indices.

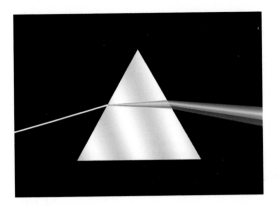

COLOR FIGURE 2.5 Prism diffraction: analyzing white light to its constituent colors.

COLOR FIGURE 2.8 The diffraction of light when incident on surfaces such as a compact disk, which consists of several grooves, leading to interference and the formation of colors.

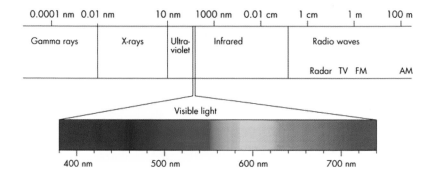

COLOR FIGURE 2.10   The electromagnetic spectrum.

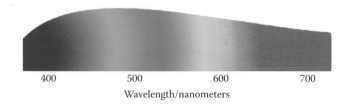

Wavelength/nanometers

COLOR FIGURE 2.15   The color analysis/spectrum of daylight.

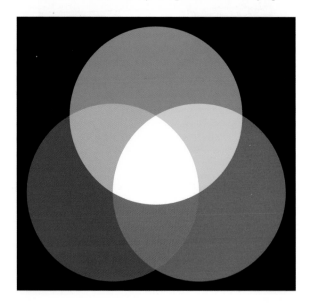

COLOR FIGURE 2.16   The combination of the three primary colors gives white light. Any color can be created with the appropriate combination of primary colors.

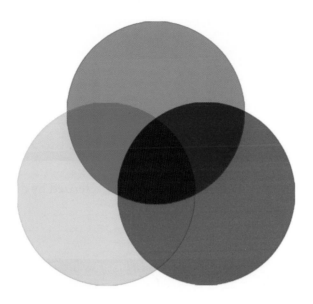

COLOR FIGURE 2.17   Media that use reflected light and its partial absorption in order to create colors are using the abstractive method of mixing colors.

| 1800 K | 4000 K | 5500 K | 8000 K | 12000 K | 16000 K |

COLOR FIGURE 2.20   Scale of color temperature.

COLOR FIGURE 2.21   Different lamps are characterized by different color temperatures and color rendering indices.

COLOR FIGURE 2.22   Different lamps are characterized by different color temperatures and color rendering indices.

| 2800 K | 3800 K | 6000 K | 9500 K |

COLOR FIGURE 2.23   Different lamps are characterized by different color temperatures and color rendering indices.

COLOR FIGURE 3.1   Visible light emissions from an incandescent material.

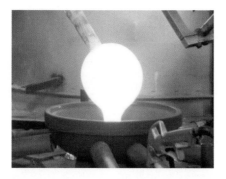

COLOR FIGURE 3.2   Visible light emissions from an incandescent material.

COLOR FIGURE 3.7   Operating incandescent lamps emitting their characteristic warm white light.

COLOR FIGURE 3.8 Operating incandescent lamps emitting their characteristic warm white light.

COLOR FIGURE 3.9 Operating incandescent lamps emitting their characteristic warm white light.

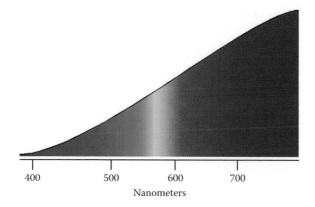

COLOR FIGURE 3.13   The emission spectrum of an incandescent lamp is a continuous one.

COLOR FIGURE 3.23   The addition of silica on the tungsten filament resulted in fluctuations of the brightness. In the left column, one sees the brightness of the tungsten filament under 20 W. In the right column, one sees the bursts of brightness increase when silica is added.

COLOR FIGURE 3.24 The addition of silica on the tungsten filament resulted in fluctuations of the brightness. In the left column, one sees the brightness of the tungsten filament under 20 W. In the right column, one sees the bursts of brightness increase when silica is added.

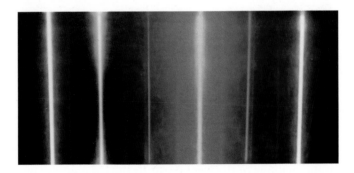

COLOR FIGURE 4.3 Electrical discharges through different gases. Each gas emits radiation at signature wavelengths. The light emitted from different plasma sources is characteristic of the gas or vapor fill.

COLOR FIGURE 4.23   Low-pressure gas discharge lamps for decoration and signs.

COLOR FIGURE 4.24   The neon discharge lamp has dominated the advertising signs industry for decades.

COLOR FIGURE 4.33   Low-pressure sodium vapor discharge lamps for street and road lighting.

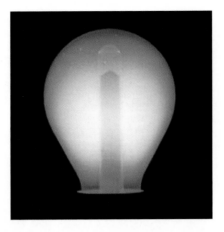

COLOR FIGURE 4.70 Photograph of an induction mercury lamp without the phosphor coating.

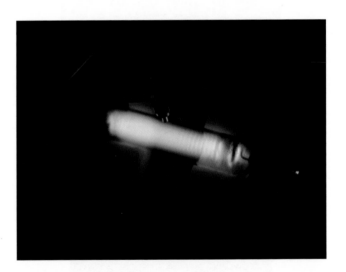

COLOR FIGURE 4.81 Photograph and emission spectrum of a low-pressure metal halide lamp (AlCl3), which has been proposed in the past as an alternative to mercury-filled low-pressure discharge lamps.

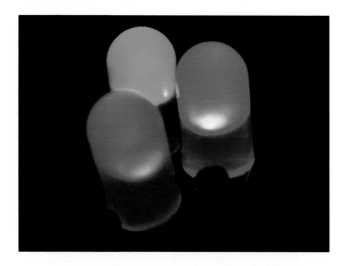

COLOR FIGURE 5.11 Combination of different color LEDs for the creation of white or dynamic lighting.

COLOR FIGURE 5.12 Combination of different color LEDs for the creation of white or dynamic lighting.

COLOR FIGURE 5.17   LEDs in series mainly for decorative lighting.

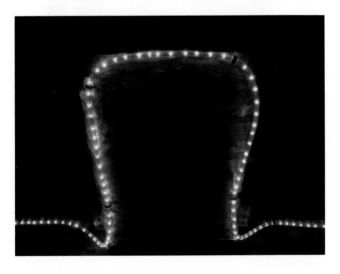

COLOR FIGURE 5.18   LEDs in series mainly for decorative lighting.

COLOR FIGURE 6.1   For the creation of all colors, one must use three light sources in a system.

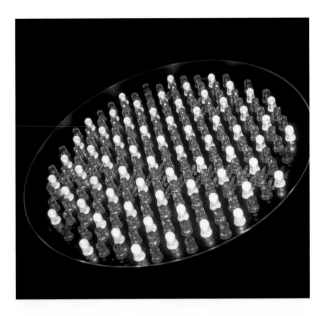

COLOR FIGURE 6.2 Dynamic lighting can be achieved with the use of multiple colored sources in the same system.

COLOR FIGURE 6.3 Dynamic lighting can be achieved with the use of multiple colored sources in the same system.

COLOR FIGURE 8.1   Laser light emissions (Nd: YAG 532 nm) in research application.

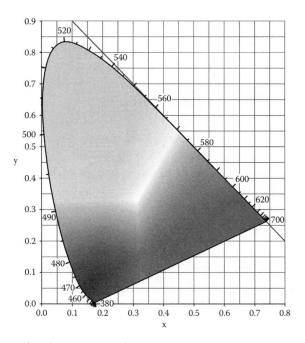

FIGURE 2.24 The chromaticity diagram.

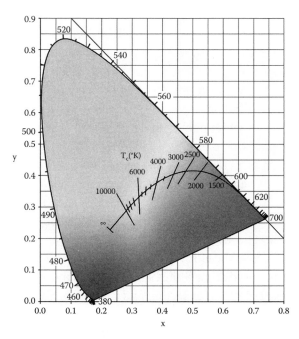

FIGURE 2.25 Chromaticity diagram with the blackbody locus and correlated color temperatures for various light sources.

2.4 WHITE LIGHT

If a source emits in the entire visible spectrum (such as an incandescent light source), the spectrum is called continuous and, of course, the light we receive is white. The color temperature of this white light will depend on the balancing between the different regions, so, for example, we are dealing with a warm white of around 3000 K if the spectrum is stronger in the red region (such as incandescent sources) or with a cool white if the blue region is quite intense (5000 K or more). In the appendices, the reader will find a scale of different color temperatures for different sources. The important thing to note is that a continuous spectrum emitting at all wavelengths of the visible spectrum will appear white and will be characterized by the highest CRI as all colors can be reproduced under such a source.

The human eye has three groups of sensors, each being responsible for detecting three different regions of the visible spectrum. The peaks of the three sensitivity curves of those sensors (maximum sensitivity) are at the blue, green, and red regions. In other words, a source that emits just three lines (or in general, three sharp and narrow emissions at various wavelengths), one at each of the regions that the different groups of sensors are sensitive to, then the end result is that the human eye will again perceive the source as white because all sensors will be "activated." The difference, of course, between such a source and a continuous one is that although they both appear white, the noncontinuous one will be characterized by a lower CRI as not all colors of surrounding objects can be reproduced.

We make use of this effect in technologies in which a continuum source is not available but white light is to be produced. The way to do it is to employ a phosphor that under an excitation emission line from the available source (such as mercury's UV emission lines in a discharge tube or an LED's UV narrow band) will produce visible emissions. The phosphor can be tailored to either produce a wide continuous spectrum if we want to have white light with high CRI; or, if the goal is efficiency, then a phosphor is chosen so that it produces only some lines, those that will stimulate our eye's sensors and give the effect of white light (but with low CRI).

The method of producing white light with phosphors is discussed both in the discharge light sources section and Chapter 5. Some of those phosphors are referred to as tri- or tetrachromatic, indicating that they are of the kind that produce only the three or four narrow emissions required to give a white light effect and high efficacy but low CRI. Any conversion of radiation using phosphors includes, of course, losses, because higher-energy photons are being converted into lower-energy ones (Stoke losses), and one of the holy grails of light source scientists is to be able to develop phosphors that do not introduce energy differences. This would be achieved if a phosphor was developed that, for each high-energy photon it absorbs, would emit more than one photon of lower energy, or if a phosphor could be excited by lower-energy photons (red or infrared), giving off a higher-energy photon that would have the sum of the energies of the stimulating ones. Such a phosphor would revolutionize the light sources field, as not only would it dramatically increase the efficacy of the sources that already make use of phosphors, but it could even dramatically increase the efficacy of lamps such as incandescent sources, where most of the radiation is in the infrared and we are currently unable to use this part for lighting applications. Perhaps this process will be achieved in future using different technologies, similar to how quantum dots (see Chapter 5) introduced a new way to convert radiation into different wavelengths. A similar concept is discussed in Chapter 3, where we discuss electrocandoluminescent technology. In this case, materials that when heated give off light of higher color temperature than blackbody radiation laws would predict, could be employed with an incandescent source (an infrared source) the same way as they used to be employed with nonelectrical sources in the past (limelight, gas mantle lamps, etc.).

Having already discussed the method for producing white light using a phosphor that produces the primary colors (the colors to which our three different eye sensors are most sensitive), we must also mention a final technique that is considered the cheapest and easiest method (as a rule, the energy-efficient ways for producing white light imply lower CRIs, as fewer parts of the visible spectrum are covered by the produced emissions).

The three curves that depict eye sensitivity to different colors overlap, so a yellow emission line (yellow lies between green and red) is able to stimulate both groups of sensors responsible for the green and red colors. In other words, a source emitting just blue and yellow light will also appear to the human brain as a white source. This source will, of course, have a very low CRI as most objects except the blue and yellow ones will not be reproduced correctly. This easy method of producing white light (because of the blue color involved, the source gives off a cool white light) is exploited in LED technology as the cheapest technique of producing white LED sources (a blue LED with a phosphor on top converting some of the blue light into yellow). In the case of discharge tubes, such a technique is also possible (mercury's UV emission at 254 nm could be converted into just blue and yellow), but to my knowledge there is no such product, and most phosphors convert the UV radiation into more colors. However, this effect has been observed and studied by the author with medium-pressure mercury discharges, where the visible lines of mercury are significantly enhanced under certain conditions, and there is no need for any phosphor (see Chapter 4).

Incandescent lamps and many high-pressure discharges emit broadly in the visible without the need for phosphors. However, other sources such as low-pressure discharges and LEDs employ the aforementioned techniques of combining specific wavelengths or using phosphors for the creation of white light, and each case will be mentioned in the relevant sections.

2.5 MEASURING LIGHT

The efficiency of a source is defined as the percentage of electrical power converted to radiation, and the luminous efficacy is defined as the percentage of power/energy converted to visible radiation. The efficacy is measured in lumens per watt, and the theoretical maximum is 683 lm/W if all the electricity is converted into radiation with a wavelength corresponding to the maximum sensitivity of the eye at 555 nm.

For measuring light from a source, the following terms are useful.

Luminous flux, Φ, is defined as the amount of light emitted per second from a light source, and the unit is the lumen (lm = cd × sr), as shown in Figure 2.26. The measurement of the lumen flux is usually done with an integrating sphere (Figure 2.27), where the light is diffusely reflected by the inner walls. Each unit surface of the sphere is illuminated proportionally to the flux of the light source, and a small window collects and measures this luminous flux.

The surface of a sphere is $4\pi r^2$, where r is the radius of the sphere, so the solid angle that is defined by the entire sphere is

$$\Omega = 4\pi \text{ (sterad)}$$

The solid angle of a surface A that is part of a notional sphere is defined as

$$\Omega = A/r^2$$

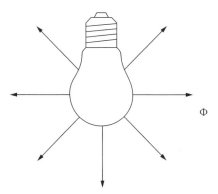

FIGURE 2.26   Luminous flux Φ.

A unit such as the lumen is defined as the total radiated power in watts times the efficacy of the light bulb times 683 lm/W.

Luminous intensity (I) is defined as the flux of light emitted per second in one direction, and its unit is the candela (cd = lm/sr). It is measured by a goniophotometer (goniometer), which is shown in Figure 2.28.

$$I = \Delta\Phi/\Delta\Omega$$

where Φ is the luminous flux and Ω is the solid angle.

FIGURE 2.27   Integrating sphere (Ullbricht sphere) for measuring lamp luminous flux.

FIGURE 2.28   Goniophotometer.

One candela is defined as the luminous intensity of a source in a particular direction that emits monochromatic light of wavelength 555 nm and power equal to 1/683 w/sr.

If a source radiates with equal intensity in all directions, we call it isotropic, and each time the distance from the source doubles, the intensity measured becomes four times lower, while if the radiation surface is flat, the intensity decreases by the cosine of the angle of measurement according to Lambert's law (as for luminance), which is schematically depicted in Figure 2.29.

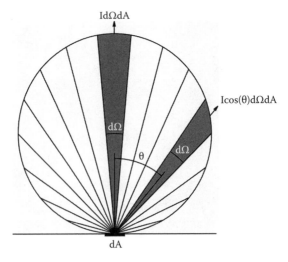

FIGURE 2.29   Graphical representation of Lambert's law.

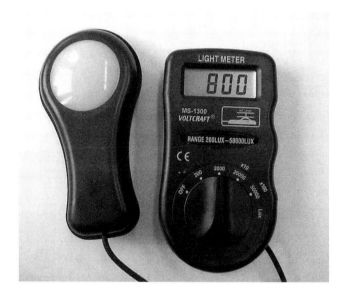

FIGURE 2.30   Commercial photometers.

Illuminance, E, is defined as the amount of light incident on a unit area, and the unit of measurement is the lux ($lx = lm/m^2$). Illuminance is measured by photometers that convert light into electricity (luxmeter). Some commercial luxmeters can be seen in Figure 2.30.

$$E = \Phi/A = I/r^2$$

where I is the luminous intensity and A is the area of incidence.

Luminance, L, has the unit cd/m^2, and is defined as the amount of light emitted from a unit surface in a particular direction.

It is also defined by the formula $L = I_f A$.

CHAPTER 3

Electrical Incandescent Light Sources

THIS CHAPTER DESCRIBES INCANDESCENT light sources, which were the first—and even today are perhaps the most important—light-producing technology in the market based on electricity.

For more than a century, incandescent lamps have been the mainstay of artificial lighting. Today, although discussions are under way related to the banning of this technology due to economic, energy, and environmental reasons, most of these lamps still play an important role.

Many European countries have started a gradual banning of the incandescent lamp. This banning will initially include conventional incandescent lamps of high wattage but not reflector lamps and spot lights such as halogen lamps. So, an evaluation of this technology is still very relevant and remains important to the light sources professional.

Incandescent light sources have a rich history of developments and discoveries, their own unique path of technological evolution, and characteristics that make this type of lamp still desirable. At the end of this chapter, some new developments and ideas are discussed that could not only bring this classic technology back in favor but even lead to a new invention and the birth of a new novel light source technology.

Incandescent lamps, as the name implies, are based on the phenomenon of incandescence. The principle does not differ from that of the blackbody, which radiates as it is being heated, starting from the infrared part of the spectrum and covering more and more of the visible spectrum as the temperature increases. Figures 3.1 and 3.2 show molten glass, which begins to radiate in the visible spectrum as the temperature increases. The

FIGURE 3.1 **(See color insert following page 20.)** Visible light emissions from an incandescent material.

heating here is provided by the electric current flowing through a solid material in the same way that the electric oven or kettle operates.

The flow of electricity through a filament increases its temperature as it resists the flow of electrons. The increase in temperature excites the electrons of the filament atoms to higher energy states, and as they return to their original states, they release this energy by emission of photons.

Due to continuous blackbody-like emissions, incandescent lamps have excellent color rendering properties and are given an index of 100. The biggest proportion of the emitted radiation is, of course, in the infrared part of the spectrum and almost 90% of the electrical energy is lost as heat. With regard to the remaining 10% of the electrical energy, most of it is converted into red emissions, giving the source a warm white color. Incandescent lamps are therefore ideal for creating warm environments and reproducing the colors of the objects illuminated. They are not, however, an economic

FIGURE 3.2 **(See color insert following page 20.)** Visible light emissions from an incandescent material.

solution since their efficacy does not exceed 20 lm/W. A simple multiplication of the total power consumed by the lamp times the efficacy number gives us the maximum luminous flux in lumens. There are also lamps with colored glass bulbs that act as filters, thus giving off light of specific colors or different tones of white. Since the colored glass blocks undesired wavelengths in order to produce the desired color or decreases the amount of red that passes, the lamp becomes even less efficient.

3.1 CONVENTIONAL INCANDESCENT LAMPS

The name that is linked with the invention of the electric lamp, and more specifically with the electric incandescent lamp, is that of Thomas Alva Edison, and 1879 (the year Edison filed his patent for a practical lamp with a carbon filament) is referred to as the birth year of this lamp. Thomas Edison was a prolific inventor, and his name is connected with the incandescent lamp to such a degree due to some important contributions that led to the development and commercialization of this product. But the history of this lamp began decades before Edison's time and is still being written today. This history includes many technologists and scientists from various countries who played important roles, while the decisive factor in the evolution and spread of the incandescent lamp was the distribution of electricity to greater parts of the population.

During the first attempts to convert electrical energy into light by means of incandescence in the early 19th century, scientists used platinum and also carbon filaments inside a glass bulb. The bulb was always under vacuum or contained a noble gas at low pressure in order to protect the filament from atmospheric oxygen, which accelerated the process of filament evaporation leading to the end of lamp life. After nearly a hundred years of testing, the incandescent lamp took its current form, which employs tungsten filaments. A photo with a detailed analysis of the anatomy of a modern conventional incandescent lamp can be seen in Figure 3.3. Figure 3.4 shows the variety of shapes and forms of incandescent lamps used today, while Figures 3.5 and 3.6 show incandescent lamps used with decorative luminaires for indoor lighting.

Here we note in brief some important milestones of the path toward maturity of the incandescent lamp and the people behind them:

1809—Humphry Davy demonstrates to the Royal Society of Britain the phenomenon of creating light through incandescence by using a platinum filament.

FIGURE 3.3   The anatomy of an incandescent lamp.

1835—James Bowman Lindsay presents an incandescent lamp in Scotland.

1840—Warren de la Rue heats a platinum filament under vacuum.

1841—Frederick de Moleyns files the first patent for an incandescent lamp with a platinum filament under vacuum.

1845—John Starr files a patent for an incandescent lamp with a carbon filament.

1850 to 1880—Joseph Wilson Swan, Charles Stearn, Edward Shepard, Henrich Globel, Henry Woodward, and Mathew Evans develop independently of each other incandescent lamps with various carbon filaments.

FIGURE 3.4   Various types of incandescent lamps.

FIGURE 3.5 Various types of incandescent lamps are used widely in house lighting luminaires.

FIGURE 3.6 Various types of incandescent lamps are used widely in house lighting luminaires.

1875—Herman Sprengel invents the mercury vacuum pump. The better vacuum-creating technique leads to better lamps.

1878—Thomas Edison files his first patents for incandescent lamps with platinum and, later, carbon filaments while he also starts distribution of the lamps.

1878—Hiram Maxim and William Sawyer start the second incandescent-lamp-selling company.

1898—Carl Auer von Welsbach files his patent for a lamp with an osmium filament. That was the first commercial lamp with a metallic filament.

1897—Walter Nerst invents the Nerst lamp.

1903—Willis Whitnew develops carbon filaments with metallic coatings, thus reducing the blackening of the glass.

1904—Alexander Just and Franjo Hanaman file the patent for an incandescent lamp with a tungsten filament.

1910—William David Coolidge improves the method for producing tungsten filaments.

1913—Irving Langmuir uses noble gases instead of vacuum in incandescent lamps thus making them more efficient and reducing glass blackening.

1924—Markin Pipkin files his patent for frosted glass.

1930—Imre Brody replaces argon with krypton or xenon.

Incandescent lamps play key roles in indoor lighting such as home lighting due to the warm and comfortable white light they emit, the variety of shapes they come in, and their simple operation. The warm light (more red than blue in the emitted continuous spectrum) emitted by incandescent lamps used in a typical home can be seen in Figures 3.7, 3.8, and 3.9.

The choice of tungsten was made for the simple reason that it is a metal with a high melting point (3660 K), so it allows for the operation of lamps at relatively high filament temperatures (2800 K for a conventional incandescent lamp). Higher temperatures mean greater efficacy because a larger

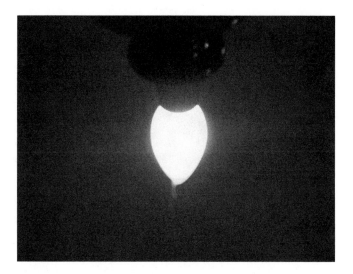

FIGURE 3.7 **(See color insert following page 20.)** Operating incandescent lamp emitting their characteristic warm white light.

FIGURE 3.8 **(See color insert following page 20.)** Operating incandescent lamp emitting their characteristic warm white light.

FIGURE 3.9 **(See color insert following page 20.)** Operating incandescent lamps emitting their characteristic warm white light.

percentage of radiation will be in the visible, but also a decrease of the average lamp life.

Greater power means greater efficacy. For a 5 W lamp with 25 lm flux, the efficacy is about 5 lm/W, while for a 250 W lamp with a flux of 4000 lm, the efficacy is 20 lm/W.

More support wires make the filament mechanically stronger but reduce the temperature to their thermal conductivity, so one has to choose between efficacy and lifetime. Other elements such as thorium, potassium, aluminum, and silicon or combinations of those may be added to improve strength.

Some lamps have two filaments with three connections at their bases. The filaments share a common ground and can be used together or separately. For these lamps, three numbers are given where the first two show the consumed power on each filament, and the third number is their sum.

To improve the function of the lamp, the filament is in spiral form, as seen in Figure 3.10, which provides economy of space and heat. The power of the lamp depends on the operating voltage and the resistance of the filament as it is heated, and the resistance of the filament depends on the length and diameter. The value of the resistance of a cold filament is about 1/15 of that under operation. For a given lamp power, voltage, and temperature, the filament has a specific diameter and length. Small differences in the resistance value along the length of the filament, which are mainly due to differences in the diameter, result in the creation of

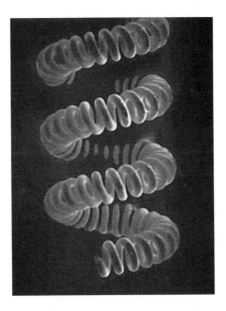

FIGURE 3.10   A tungsten spiral filament.

hot spots. Tungsten is evaporated faster at these hot spots, leading to a dramatic decrease of the lifetime of the lamp. Figures 3.11 and 3.12 show incandescent filaments emitting the characteristic red glow or warm white light.

There are no specifications for the glass of the bulb, but the supports of the filament are made of molybdenum and copper so that they melt in case

FIGURE 3.11   Incandescent tungsten filament.

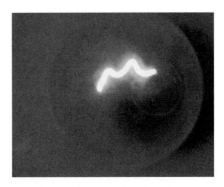

FIGURE 3.12   Incandescent tungsten filament.

of high current, thus acting as fuses, and they are enclosed in lead glass, which can provide the necessary insulation. The bases of the lamps are usually screw bases, known as Edison bases, having the code letter R followed by a number indicating the diameter in millimeters, or Bayonet bases, also having a number indicating the diameter and the code letter B.

In case the bulb is not under vacuum, the gases contained are argon and nitrogen at low pressure. The main purpose of the inclusion of the gas is to reduce or control the burning of the filament, and a more specific reason for using nitrogen is to prevent formation of sparks.

If there is a leak in the bulb, then the filament comes in contact with atmospheric air, producing tungsten oxides that solidify on the walls. If the bulb is under vacuum, then the blackening of the walls by the dark tungsten oxides is homogenous. In case the bulb contains a gas such as argon, then the oxides are transferred to the top of the bulb due to convection currents. The blackening of the bulb is the second most important factor, after filament vaporization, in reducing the light output. Water is probably the impurity that mostly leads to wall blackening, and the way it contributes toward that has been named the water cycle. Inside the bulb, the water molecules break up, creating molecules of oxygen and hydrogen. The oxygen molecules react with tungsten, and the tungsten oxides that result travel toward the cold spots of the bulb, where they deposit. The hydrogen molecules reach those spots and react with the oxides, resulting in tungsten being left on the wall and new water molecules ready to start a new cycle.

Incandescent lamps are available in many shapes and sizes, while the applied voltage can be from 1.5 to 300 V. Some of the important advantages of the incandescent lamps are their low manufacturing cost, and they operate with both DC and AC current.

Another advantage of this type of lamp is that it functions as a simple resistor, so the connection voltage and current are proportional. This means that no additional gear, such as the ballast found with fluorescent lamps, is needed. The lamp accepts the voltage of any household outlet, except in the case of some special low-power lamps for which the voltage is limited to 12 V by means of an integrated transformer. Not using a transformer certainly has advantages, but the application of high voltage would mean that in order to keep the power constant the current would have to be reduced, so the filament would have to be made thinner and longer. These changes make both the lamp and the halogen cycle less efficient.

The emitted spectrum of these lamps is continuous and covers the entire visible range, as shown in Figure 3.13. This means that such light sources have a very good color rendering index, and it has been designated as 100. The largest percentage of radiation is in the infrared, so around 90% of the electric energy is lost in the form of heat. Regarding the 10% of visible radiation emitted, most of it is in red, giving the white color of the source a warm tone.

Incandescent lamps are therefore ideal for creating cozy environments and for applications where very good reproduction of colors is needed. However, they are not an economic solution, since the efficacy is small (not exceeding 20 lm/W) and the average lifetime is around 1000 h. A simple multiplication of the total power of the lamp by the efficacy value gives the luminous flux in lumens.

Average rated lifetime is defined as the time duration beyond which, from an initially large number of lamps under the same construction and under controlled conditions, only 50% still function. Measurements of

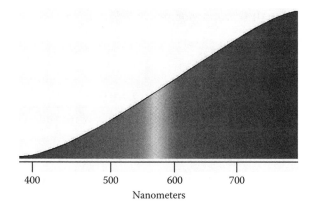

FIGURE 3.13 **(See color insert following page 20.)** The emission spectrum of an incandescent lamp is a continuous one.

rated average lamp lifetimes are usually made by applying an operating cycle. For example, the lamps can be operated for 18 h a day and remain switched off for the other 6 h or 3 h on and 1 h off.

Such measurements offer a good basis for comparisons on technical life and reliability, although the same figures are unlikely to be obtained in practice, because parameters such as supply voltage, operating temperature, absence of vibration, switching cycle, etc., will always be different.

The *service lamp life* is another term and depends on the lifetime and lumens maintenance. Often, 70% service life or 80% service lifetime is used. This is the number of operating hours after which, by a combination of lamp failure and lumen reduction, the light level of an installation has dropped to 70% or 80% compared to the initial value.

Attention must, however, be paid to changes in the operating voltage because an increase in voltage will increase the brightness but will also reduce the lifetime, and vice versa. Many "long life" lamps are based on exactly this relationship.

The temperature of the filament depends on the voltage, while the average life, current intensity, power, luminous flux, and efficiency depend on the filament temperature. Thus, all parameters have a relationship with voltage, which has been calculated as follows:

Current intensity (A) proportional to $V^{0.48-0.62}$

Mean lifetime (h) proportional to $V^{11.8-14.5}$

Power (W) proportional to $V^{1.48-1.62}$

Luminous flux (lm) proportional to $V^{3.3-4}$

Efficacy (lm/W) proportional to $V^{1.84-1.93}$

3.2 HALOGEN INCANDESCENT LAMPS

The halogen lamp is based on the same principle of incandescence, but the advantage it offers, compared to the ordinary/conventional incandescent lamp, is that its *rated average lifetime* is twice as long, exceeding 2000 h.

The increase in the average lifetime of the halogen lamp compared to a conventional incandescent lamp is the result of a chemical balance that takes place within the lamp between the tungsten filament and halogen gas. The chemical balance is called the *halogen cycle*, and the gas used is bromine (or compounds of bromine), which replaced iodine, which was used in the first halogen lamps. It is because of the inclusion of iodine in the early lamps that the term iodine lamp is still used today.

The halogen cycle works as follows: A tungsten atom that has escaped to the gas phase, due to the high temperature of the filament, interacts and bonds with a halogen atom, forming a compound that is not deposited at the cold point of the lamp, but continues to travel in the gaseous phase until it again reaches the filament, where the high temperature dissociates the compound; the tungsten atom returns to the filament and the halogen atom to the gas phase, ready to start a new cycle (Figure 3.14).

This cycle allows operation of the lamp at higher temperatures (3000–3500 K compared to 2800 K for conventional lamps) with higher luminous efficacy, intensity, and color temperature. Because of the higher temperatures required for the halogen cycle to take place, a harder glass such as quartz is used, and the dimensions of the lamp are smaller while the filament is thicker. By using a stronger glass, an increase of the gas pressure is also allowed, reducing further the vaporization of tungsten. Quartz is transparent to UV radiation below 300 nm (the usual soft glass is not transparent to electromagnetic waves with wavelengths shorter than 300 nm), so an additional filter is required if this radiation is not desirable. The additional filter also provides protection from glass fragments in case the bulb shatters.

Another method of blocking ultraviolet radiation is by using specially doped quartz glass. A halogen lamp with no filter or special quartz glass can be used as a source of ultraviolet radiation in special scientific applications.

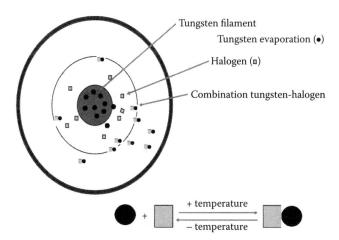

FIGURE 3.14 The halogen cycle.

Atmospheres of tungsten-halogen lamps comprise an inert gas with about 0.1% to 1.0% of a halogen vapor added. The inert gas may be xenon, krypton, argon, or nitrogen, or a mixture of them having the highest atomic weight consistent with cost as well as arc resistance suitable to the lamp design and the operating voltage. The halogen vapor may be pure iodine (I_2) or a compound of iodine (e.g., CH_3I) or of bromine (e.g., HBr, CH_3Br, or CH_2Br_2). The minimum bulb wall temperature for operation of the halogen cycle is about 200°C for bromine, which is significantly lower than 250°C for iodine. Bromine is also colorless, while iodine has a very slight absorption in the yellow-green.

Unlike conventional tungsten-filament lamps, which operate with an internal gas pressure of about one atmosphere, most tungsten-halogen lamps operate with an internal gas pressure of several atmospheres to reduce the rate of tungsten evaporation.

The amount of halogen that is added is such that it balances the evaporation rate of tungsten at the nominal voltage. An increase in the voltage leads to higher rates of tungsten evaporation, rendering the amount of halogen insufficient and leading to wall blackening. If the applied voltage is lower than the nominal one, then the lamp temperature might be too low for the halogen cycle to work. At least in this second case, the evaporation rate of tungsten is low enough for blackening not to occur.

The halogen lamps compared to ordinary/conventional incandescent lamps have longer lifetimes (2000 h or more compared to 1000 h), higher color temperatures (3000–3500 K compared to 2800 K), and higher luminous efficacies (30 lm/W compared to 20 lm/W). A typical halogen incandescent lamp is shown in Figure 3.15.

Attention should be paid to avoid impurities on the surface of the lamp, especially fingerprints, which can cause damage to quartz under the high-temperature conditions.

Lamps with reflectors (reflector and parabolic aluminized reflector) can be used as projection lamps for directional lighting (Figures 3.16–3.19). Besides the known bases, halogen lamps have bases with a double contact carrying the code letter G followed by the size of the base in millimeters.

All incandescent lamps produce large amounts of heat, which is emitted together with the visible light toward the desired direction after reflecting from the aluminum reflector of each lamp. If the aluminum reflector is replaced by a *dichroic* reflector, then we talk about a cool-beam lamp since at least 2/3 of the infrared radiation passes through the reflector and only the visible radiation and the remaining 1/3 of the reflected infrared

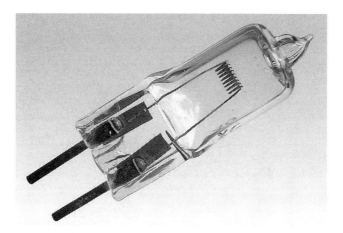

FIGURE 3.15   Halogen lamp.

radiation pass through the glass and are emitted in the desired direction. Therefore, the materials behind a cool-beam lamp, such as the base of the lamp where the largest proportion of the infrared radiation escapes to, should be able to withstand this exposure to thermal radiation.

Halogen lamps offer a compromise between conventional incandescent lamps and more efficient compact fluorescent lamps. Halogen lamps still have all the advantages offered by the incandescent technology such as

FIGURE 3.16   Halogen lamp with reflectors.

FIGURE 3.17   Halogen lamp with reflectors.

FIGURE 3.18   Halogen lamp with reflectors.

FIGURE 3.19   Halogen lamp with reflectors.

TABLE 3.1 Various Parameter Values for Various Lamp Technologies

	Efficacy lm/W	Power/W	Color Rendering Index	Average Lifetime/Hours
Incandescence	20	15–1,000	100	1,000
Halogen	30	5–2,000	100	2,000–5,000
Fluorescent	55–120	5–125	55–99	10,000–25,000
Inductive mercury	70–80	55–165	80	60,000–100,000
Sodium low pressure	200	35–180	0	20,000
Xenon DBD	30	20–150	85	100,000
Xenon high pressure	30	1,000–15,000	90	2,000
Sodium high pressure	50–150	35–1,000	25–85	10,000–30,000
High-pressure mercury	60	50–1,000	15–55	10,000–30,000
Very high-pressure mercury	60	100–200	60	10,000
Metal halide	70–100	35–2,000	70–90	10,000–20,000
Sulfur	95	1,500	80	60,000 (20,000 driver)
LEDs	30	0.1–7	0–95	50,000–100,000

instant start and peak brightness, complete control of dimming ratios, absence of harmful materials such as mercury, lifetime independent of switching frequency, etc.

On the other hand, they offer additional advantages compared to conventional incandescent lamps such as longer lifetimes for the same or greater efficacy and a higher color temperature. See Table 3.1 for a complete comparison of various lamp characteristics between different lamp technologies.

Another development that further increases the luminous efficacy of halogen lamps is the *Infrared Reflecting Coating (IRC)*. The inside wall of the lamp is covered with several dichroic coatings that allow the transmission of visible light but reflect part of the infrared radiation back to the filament, raising the temperature to the desired levels with less energy consumption. The efficiency increases by up to 40% compared to conventional halogen lamps and the average lifetime reaches up to 5000 h.

Halogen lamps behave in a similar way as conventional incandescent lamps when the applied voltage differs from the specified one. Small increases of voltage lead to increases of luminous flux and efficacy but also to decreases of the average lifetime. Each parameter has a different proportionality value, but approximately a 10% increase in the voltage value leads to a similar increase of the flux and efficacy and a 50% decrease of the lifetime.

In general, for the optimum operation of the lamp, it is best to avoid changes with regard to the specified values by the manufacturer, especially when it comes to the applied voltage.

Attention should be paid to avoiding impurities on the bulb surface, particularly fingerprints, as they damage the quartz glass at high temperatures. To remove the impurities, the bulb must be cleaned with ethanol and dried before use.

All incandescent lamps produce large amounts of heat that is emitted along with the visible light in the chosen direction after reflecting off the aluminum reflector found in each lamp. If the aluminum reflector is replaced by a dichroic one, then we can have what we call a cool beam, since at least 2/3 of the infrared radiation pass through the reflector and only the remaining 1/3 and the visible light is reflected back toward the desired direction. In such cases, the materials behind a cool-beam lamp such as the base, where most of the infrared radiation escapes to, must be able to cope with this exposure to thermal radiation.

Of course, the production of heat by incandescent lamps is not desirable in lighting, but in other applications where heat is required, these lamps have a significant advantage over other light source technologies. For example, proponents of the incandescent lamp claim that heat produced balances some of the energy that would otherwise be needed to heat a place in the winter.

The low cost of this kind of lamp is also a major advantage and must be taken into account when considering the phaseout of this technology, as it is the only one affordable for a large percentage of the world population. In poor communities, the introduction of other technologies would only be possible under strong or total financial coverage by governmental or private institutions.

Incandescent lamps have many applications in houses, shops, restaurants, and offices since they offer very good color rendering, and their variety of bulb geometries allows them to be used for decorative lighting.

3.3 EVALUATION OF THE TECHNOLOGY

Table 3.2 shows some important lamp characteristics and which of the main three light source technologies offers them to the greatest degree. Although some important characteristics such as efficacy and lifetime are offered to a greater degree by discharge lamps (fluorescent, neon, sodium, etc.) and solid-state lighting (light-emitting diodes), incandescent lamps offer advantages in almost all other cases.

Incandescent lamps offer instant start and peak brightness, total control of dimming ratios, absence of any harmful material such as mercury, and the fact that switching frequency does not affect their lifetimes.

Other advantages of incandescent lamps are the low manufacturing cost, their ability to operate with both alternating and direct current, and the fact that extra gear is not required.

Their color rendering index is exceptional, and this kind of lamp technology has found many applications such as the lighting of homes, restaurants, shops, and offices, while their variety of shapes and forms renders them ideal for decorative lighting. Lamps with reflectors (reflector and parabolic aluminized reflector) can also be used as spot lights.

Although this technology of electrical light sources is the oldest one that penetrated the market, some of its characteristics remain unsurpassed by the other two technologies (discharges and SSL) and offers advantages that enable it to resist new developments and still play a prominent role.

TABLE 3.2   Comparison of Incandescent Lamp Technology with the other Two Main Technologies of Plasma- and Diode-Based Lamps

	Incandescent-Halogen	Electrical Discharges	LEDs
Color rendering index	√		
Range of color temperatures		√	
Instant start	√		
Lifetime			√
Switching frequency	√		
Efficacy		√	
Cost	√		
Dimming	√		
Operation (AC–DC)	√		
Absence of extra gear	√		
Toxic/harmful materials	√		
Variety of shapes—forms	√		
Range of power—voltage	√		

Research on incandescent lamps continues, with most of the work focused on the filament, as it is not only the source of light but also the key element that defines the lifetime of the source. It is expected, therefore, that novel filaments might lead to a new generation of incandescent lamps with significantly longer lifetimes and efficacies. The research is focusing currently on nanomaterials and, specifically, the use of carbon nanotubes, in the hope that the electrical characteristics of these new materials will offer advantages in a light source. Ideally, the material chosen will be one that manages to convert some of the heat produced into visible light and will also offer mechanical strength and durability. Here are some recent published papers on the subject.

Probing Planck's Law with Incandescent Light Emission from a Single Carbon Nanotube

Y. Fan, S.B. Singer, R. Bergstrom, B.C. Regan—*Physical Review Letters*, 2009—APS.

Brighter Light Sources from Black Metal: Significant Increase in Emission Efficiency of Incandescent Light Sources

A.Y. Vorobyev, V.S. Makin, C. Guo—*Physical Review Letters*, 2009—APS.

Fast High-Temperature Response of Carbon Nanotube Film and Its Application as an Incandescent Display

P. Liu, L. Liu, Y. Wei, K. Liu, Z. Chen, K. Jiang, Q...—*Advanced Materials*, 2009—interscience.wiley.com.

Carbon Nanotube Films Show Potential as Very Fast Incandescent Displays

http://www.nanowerk.com/spotlight/spotid=10776.php.

3.4 CODES/PRODUCT CATEGORIZATION

Incandescent lamps are described by various codes that give information regarding their type, base, and filament, as shown in Figure 3.20.

Type of lamp: A letter (or more) indicates the shape of the lamp, and a number (or more) shows the maximum diameter of the bulb in eighths of an inch.

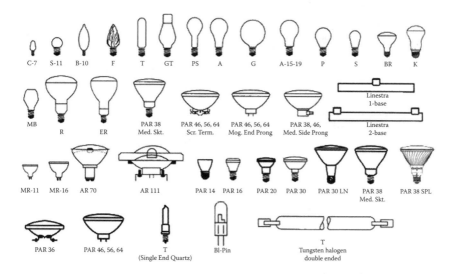

FIGURE 3.20   Codes of various types of incandescent lamps.

Filament type: The letters in front of the code indicate whether the filament is **S**traight, **C**oiled, or a **C**oiled **C**oil, while the number indicates the arrangement of the support wires. Figure 3.21 shows various types of incandescent lamp filaments.

Bases: A letter indicates the kind of base (Edison, Bayonet, G, or GY for Pins), and a number shows the maximum diameter of the base or the distance between the pins in eighths of an inch. Figure 3.22 shows various types of incandescent lamp bases.

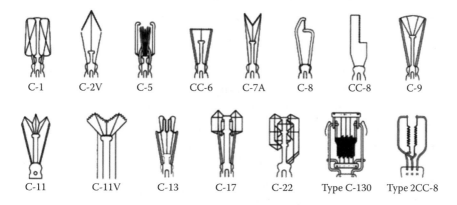

FIGURE 3.21   Types of incandescent lamp filaments.

Mini-Can Screw Mini Can (E11)   Candelabra Cand. (E12)   Intermediate Inter. (E17)   Mogul Prefocus Mog. Pt.   Single Contact Bayonet Candelabra S.C.Bay (BA15S)   Double Contact Bayonet Candelabra D.C.Bay (BA15O)   Candelabra Prefocus D.C.Pf. (P30d) S.C. Pf.   Mini Screw M.S. (E10)

Medium Skirted Med. Skt. (E26/50×39)   Medium Prefocus Med. Pt. (P28)   Medium Med. (E26)   3 Kon-Tact Medium 3 C Med. (E26D)   Screw Terminal Scr. Term   Comb, Screw Terminals & Spade Connector   Ext. Mog. End Prong   Mogul End Prong Mog E. Pr.   Base RSC

FIGURE 3.22   Types of incandescent lamp bases.

3.5 ELECTROCANDOLUMINESCENCE: A NOVEL LIGHT SOURCE

Candoluminescence is the term used to describe the light given off by certain materials that have been heated to incandescence and emit light at shorter wavelengths than would be expected for a typical blackbody radiator. The phenomenon is noted in certain transition metal and rare earth metal oxide materials (ceramics) such as zinc oxide and cerium oxide or thorium dioxide, where some of the light from incandescence causes fluorescence of the material. The cause may also be direct thermal excitation of metal ions in the material.

The emission of a bright white light from CaO when heated by an oxygen/hydrogen flame was known as limelight, and its most frequent use was in theatres and during shows to illuminate the stage and performers.

The emission of light at energy higher than expected from a flame is also known as *thermoluminescence*. Some mineral substances such as fluorite (CaF_2) store energy when exposed to ultraviolet or other ionizing radiation. This energy is released in the form of light when the mineral is heated; the phenomenon is different from blackbody radiation.

The most frequently seen case where a substance emits bright white light even under the heating of a flame that does not exceed 2500°C is the case of silicon dioxide (fused silica). It is possible perhaps to combine this luminescence phenomenon with the incandescent heating of a simple electric lamp and in a sense integrate the archaic with the modern way of lighting. The fusion of silica and tungsten could provide the means of generating bright white light even if the laws that govern incandescence and blackbody radiation would allow only for a warm/yellowish emission of light with low efficiency.

The author has recreated the effect of the bright white luminescence of silica (Figures 3.23 and 3.24) using electric heating instead of a flame, assuming that a glowing filament is at a temperature similar to that of hydrogen or methane mixed with oxygen (methane [natural gas] in air 1950°C, hydrogen in air 2055°C, propane with air 1995°C).

In the case of a hydrogen–oxygen mixture, a maximum temperature of about 2800°C is achieved with a pure stoichiometric mixture, about 700°C hotter than a hydrogen flame in air.

In the case of an incandescent lamp, the filament is heated to about 2800°C (giving blackbody light emission of the same color temperature), and in some cases such as the halogen lamp, the temperature can be raised to 3100°C. Therefore, both the electrically heated filament and a blow torch reach similar temperatures.

The author prepared lamps with silicon attached to the filament and operated them under DC mode at 20 W. For the first time, the emission of bright light was observed and recorded (see Figures 3.22 and 3.23) from the silica which was inserted inside the filament coil spiral in order to

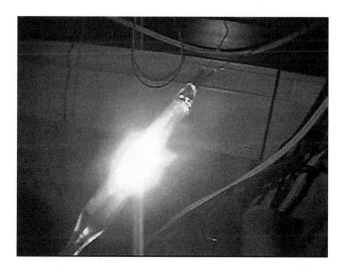

FIGURE 3.23 **(See color insert following page 20.)** The addition of silica on the tungsten filament resulted in fluctuations of the brightness. In the top column, one sees the brightness of the tungsten filament under 20 W. In the bottom column, one sees the bursts of brightness increase when silica is added.

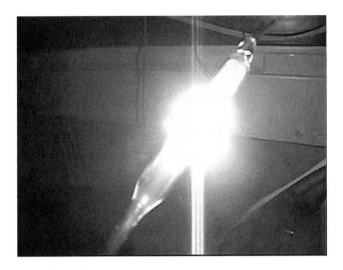

FIGURE 3.24 **(See color insert following page 20.)** The addition of silica on the tungsten filament resulted in fluctuations of the brightness. In the left column, one sees the brightness of the tungsten filament under 20 W. In the right column, one sees the bursts of brightness increase when silica is added.

ensure good contact between them and reduction of heat losses. The filament glows with a warm white light after 15 W of power input, and at 20 W, which was the maximum power input, the source fluctuates with short bursts of bright white light of higher color temperature (colder white). This was the effect of the silica reaching the right temperature for the luminescence effect to appear. The lamps gave us for the first time a demonstration of this luminescence effect, previously seen only under flame heating.

Since silicon, at a temperature of less than 3000°C, emits light of significantly higher color temperature, it could be possible to design an incandescent lamp that could offer, using this effect, light emission of much higher color temperature and higher efficacy. Some problems to be addressed are the survival of the filament and the maintenance of silica near the hot spots.

The many advantages that incandescent lamps offer, such as the simplicity of their operation and the low cost, will probably keep them in the market for a while longer. However, if they want to compete with other existing and future technologies, especially in the efficiency game, then the incandescent lamp must undergo some major developments.

Electrocandoluminescence or, in general, the use of materials capable of converting infrared energy to visible light, is perhaps one way to go. Potentially more research on filament exotic materials that emit more visible light under heating could be another way. In any case, the incandescent lamp has won rightfully its place in history as one of the most important inventions of humankind, continues to serve us, and light source researchers could be able to reinvent and immortalize it.

GENERAL SOURCES

Coaton, J.R. and Marsden, A.M., *Lamps and Lighting.* (First published in Great Britain in 1997 by Arnold, a member of the Hodder Group. Co-published in America by John Wiley & Sons Inc.)

Houston, E. J. and A.E. Kennely, *Electric Incandescent Lighting,* The W. J. Johnston Company, New York, 1896.

Friedel, R. and Israel, P. *Edison's Electric Light: Biography of an Invention.* Rutgers University Press, New Brunswick, New Jersey.

Kaufman, J. (ed.), *IES Lighting Handbook 1981 Reference Volume,* Illuminating Engineering Society of North America, New York, 1981.

Waymouth, J. *Electric Discharge Lamps.* MIT Press, 1721.

http://www.gelighting.com/na/business_lighting/education_resources/glossary.htm. GE Lighting Glossary.

http://inventors.about.com/od/lstartinventions/a/lighting.htm. *History of Lighting and Lamps.*

http://freespace.virgin.net/tom.baldwin/bulbguide.html.

http://invsee.asu.edu/Modules/lightbulb/meathist.htm.

Riseberg, L., Candoluminescent electric light source, U.S. patent number 4,539,505. 1983.

Stark, D. and Chen, A., High efficiency light source utilizing co-generating sources, U.S. patent number 6,268,685. 1997.

CHAPTER 4

Electrical Discharge Lamps

4.1 PLASMA PROCESSES

This chapter is dedicated to the technology of electrical discharges or plasmas for lighting applications. Plasma, which is often called the *fourth state of matter*, is an ionized gas with extremely interesting and useful properties. The term *plasma* was first used to describe an ionized gas by Irving Langmuir in 1927 after the way the electrified fluid carrying high-velocity electrons, ions, and impurities reminded him of the way blood plasma carried red and white corpuscles and germs.

The plasma state here on earth is quite unusual and exotic, but it becomes very common in the outer layers of the atmosphere. First, the development of radio led to the discovery of the ionosphere, the natural plasma roof above the atmosphere, which bounces back radio waves and sometimes absorbs them.

It was also recognized by astrophysicists that much of the universe consists of plasmas such as our own sun (Figures 4.1 and 4.2), which is a glowing ball of hot hydrogen plasma.

Closer to home, some people have witnessed a natural plasma in the form of the auroras, a polar light display caused by the bombardment of the atmospheric constituents by high-energy particles, and surely everyone is familiar with the blinding flash of a lightning bolt.

It was not long before humans appreciated the interesting properties of ionized gases and, today, artificial plasmas are being employed in many industrial applications. One of the most important plasma properties is the emission of light. Photons in a plasma can be produced either by electron impact excitation of atoms or by recombination of charged particles.

FIGURE 4.1 Most of the light that shines through the universe originates from plasma sources.

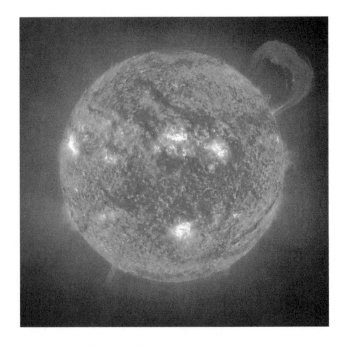

FIGURE 4.2 Most of the light that shines through the universe originates from plasma sources.

This interesting property had been identified as early as 1675 by the French astronomer Jean Picard, when he observed a faint glow in a mercury barometer tube. The glow was called the *barometric light*, and its cause was static electricity. The first development in electric lighting was the arc lamp, which was evolved from the carbon arc lamp demonstrated in 1801 by Sir Humphrey Davy, in which an electric current bridges a gap between two carbon rods and forms a bright discharge called an *arc*. By 1855, there was the Geissler tube, named after a German glassblower, where a low-pressure gas was employed, and a voltage was applied across causing the gas to glow. The neon lamp was developed by the French physicist George Claude in 1911 and became very popular due to its intense red emissions.

The beginning of the 20th century saw a rapid growth in the development and production of electric discharge lamps, but it was not until 1938 that the first practical hot-cathode, low-voltage fluorescent lamp was marketed.

The basic principle is the fact that each atom at ground energy state, if excited—which means that a valence electron leaps to a more energetic state—will return to the ground state by releasing the excess energy in the form of electromagnetic radiation. The photons emitted will have energy equal to the one that initially excited the ground atom, that is, equal to the energy state difference (the energy difference between the two orbits that the electron occupied before and after its quantum leap). Such an excitation can take place during collisions with other electrons, as in a discharge tube. Each gas emits radiation of different wavelengths, as the atomic structure of each element differs (Figure 4.3).

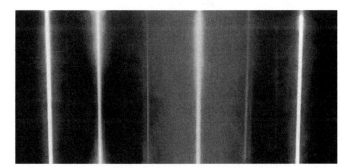

FIGURE 4.3 **(See color insert following page 20.)** Electrical discharges through different gases. Each gas emits radiation at signature wavelengths. The light emitted from different plasma sources is characteristic of the gas or vapor fill.

4.1.1 Elastic Collisions

When particles interact in the plasma medium, momentum and energy must be conserved. One could classify the collisions taking place into three different categories. The first kind is elastic collisions, where momentum is redistributed between particles and the total kinetic energy remains unchanged:

$$e^-_{fast} + A_{slow} \rightarrow e^-_{less\ fast} + A_{less\ slow}$$

It should be noted that the maximum energy a lighter particle can lose through an elastic collision with a heavier atom is $2m/M$, where m is the mass of the light particle and M is the mass of the heavier one. The efficiency of this energy transfer affects the neutral gas temperature and, consequently, the cold spot of the discharge.

For example, electrons lose their kinetic energy to atoms more efficiently via collisions with helium atoms rather than xenon. This is one of the reasons argon is used in mercury fluorescent lamps, where minimum elastic loss is desired, but neon is used in sodium lamps, where higher temperatures are required for sodium to vaporize.

4.1.2 Inelastic Collisions

The second kind of collision between particles is the inelastic type, where momentum is redistributed between particles, but a fraction of the kinetic energy is transferred to the internal energy in one or more of the particles. Collisions of this kind include electron impact excitation and ionization.

$$e^-_{fast} + A \rightarrow e^-_{slower} + A^*$$
$$\rightarrow e^-_{slower} + A^+ + e^-$$

Light particles such as electrons can lose virtually all their kinetic energy through inelastic collisions with heavier particles.

4.1.3 Superelastic Collisions

Finally, a third kind of collision can take place in which the internal energy of one of the colliding particles is transferred as kinetic energy to the other particle, resulting in an increase of the overall kinetic energy of the pair:

$$A^*_{slow} + B_{slow} \rightarrow A_{faster} + B_{faster}$$

The collision cross sections, which govern the encounter frequency and particle mean free paths, are energy dependent. For example, high-energy electrons can travel so quickly that the chances of interacting with the outer-shell electrons of an atom are reduced.

Some other plasma-phase reactions involving charged particles are the following.

$e^- + A^* \rightarrow A^+ + e^- + e^-$	Two-step ionization
$e^- + AB \rightarrow A + B + e^-$	Fragmentation
$\rightarrow A^+ + e^- + B + e^-$	Dissociative ionization
$e^- + A^+ + B \rightarrow A + B + h\nu$	Volume recombination
$e^- + AB^+ \rightarrow A + B + h\nu$	Radiative recombination
$e^- + AB \rightarrow AB^-$	Attachment
$A^+ + B \rightarrow B^+ + A$	Charge exchange
$A^* + B \rightarrow A + B^+ + e^-$	Penning ionization
$A^* + B \rightarrow A + B^*$	Energy transfer
$A^* + A^* \rightarrow A + A^+ + e^-$	Energy pooling

The excited states A^* are metastable states with lifetimes that can last up to milliseconds. Other excited states are not considered since their transition rates are very high (lifetimes of a few nanoseconds).

The parameter that characterizes a collision process is the collision cross section. Electron collision cross sections depend on impact energy as well as the scattering angles. Experimental cross sections are difficult to measure, so semiempirical formulas have been developed for different processes such as excitations and ionizations for various plasmas with mercury–argon systems being quite common due to their commercial value.

4.2 LOW-PRESSURE DISCHARGE LAMPS

4.2.1 Low-Pressure Mercury Vapor Discharges

The low-pressure mercury vapor discharge has dominated the lighting market over the past few decades, and, with efficacies reaching 120 lm/W, it is only surpassed by the low-pressure sodium discharge (up to 200 lm/W, depending on lamp wattage) used mainly in street lighting applications due to its low color rendering index (CRI). There are several billions of low-pressure mercury lamps in existence all over the world and, in the market, they represent amounts up to billions of euros annually. The active medium, that is, the element that emits radiation at a low-pressure discharge tube, such as the fluorescent lamp, is mercury, which is a metal in liquid form at room temperature. Mercury, despite its toxicity, is the most commonly used active medium in gas discharges due to its low-ionization potential (10.4 eV) and high vapor pressure. Under the operating conditions of the lamp, the pressure of the active medium depends on the temperature of the coldest spot of the lamp, which in this case does not exceed 40°C. At this temperature, the pressure of mercury is equal to 10^{-5} of the atmospheric pressure, or 7 millitorr, when the atmospheric pressure is 760 torr (1 atm = 760 torr = 760 mmHg = 101,300 Pa = 1.013 bar).

The passage of electricity through the tube vaporizes mercury, and electron collisions with the mercury atoms result in the production of ultraviolet light. This mercury vapor electric lamp, first devised by the American inventor Peter Cooper, but without the phosphor, is the most efficient way of generating light to date. The discharge converts around 60% of the electrical energy into mercury's resonance radiation at the far-UV (253.7 and to a lesser extent 185 nm), which is converted to visible light by a phosphor coated on the inner wall of the glass envelope [1,2]. It is this efficiency that has made the fluorescent lamp one of the most widely used light sources in the world, creating a huge market. The lamp also contains argon to act as the buffer gas at a pressure of 3–5 torr. The role of the buffer gas is to reduce electrode sputtering, prevent charged species from reaching the walls, and provide easier breakdown at a lower striking voltage.

Mixtures such as mercury and argon are called *Penning mixtures* due to the proximity of one component's metastable levels (argon's metastable levels at 11.53 and 11.72 eV) and the other's ionization potential (mercury's ionization potential at 10.44 eV). Breakdown can occur at lower electric fields for Penning mixtures due to the energy transfer processes that take place. Energy can be stored in the metastable levels of one component and then transferred to an atom of the other component, resulting in ionization.

4.2.1.1 The Phosphors

Under the normal conditions of operation (from mains frequency AC when electromagnetic ballasts are used to a few tens of kilohertz with electronic gear), the most intense emission line of mercury is at 254 nm, which is also known as the resonance line, and it is the result of an electron transition from the first atomic excited state to the ground state. The process of converting the UV radiation into visible light through a phosphor/fluorescent powder introduces an energy loss mechanism known as *Stokes losses,* which is not only associated with mercury but with any active medium whose emission photons have to be converted to longer wavelengths with the use of a phosphor powder. In nonluminescent materials, the electronic energy of an atomic excited state is converted completely into vibrational

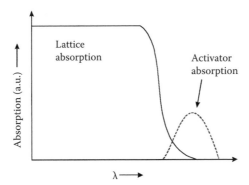

FIGURE 4.4 The Stokes losses are minimized if the phosphor excitation source is an emission line that lies close to the visible region. Short-wavelength emissions are absorbed by the lattice and converted into vibrational energy.

energy, so it is lost in the form of heat. In phosphors, the excited state returns to the ground state under emission of radiation, which is almost always at longer wavelength than the excitation wavelength. This is known as Stoke's law and, the shorter the excitation wavelength, the larger the energy loss to lattice absorptions (Figure 4.4).

As shown in Figure 4.5, the emission spectrum of a low-pressure mercury vapor discharge tube is not only linear (as opposed to the continuum spectrum), but the largest percentage of the emitted radiation is in the ultraviolet, namely, at 254 nm (see Figure 4.6 for a simplified energy level diagram of mercury).

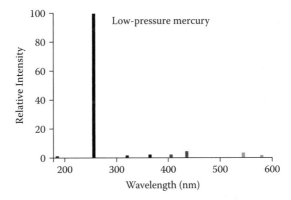

FIGURE 4.5 The emission spectrum of a low-pressure mercury vapor tube under electrical discharging.

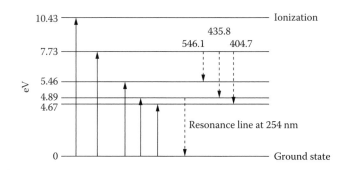

FIGURE 4.6 Simplified energy diagram of the mercury atom showing the transitions that give rise to mercury's most intense emission lines under a steady-state operation.

A large number of fluorescent powders have been developed for a large number of applications and technologies. In the case of the discharge lamp, various powders are listed in the appendices.

Depending on the application, the appropriate powder is employed, bearing in mind that improving one characteristic is usually at the expense of another characteristic, as is the case with CRI and efficiency. For example, if color reproduction is important, then a phosphor is chosen that will cover a large part of the visible spectrum, such as those listed as broadband/deluxe powders, so that the bulb has a good CRI (some phosphors can give the lamp a CRI value >95 CRI). If, however, the luminous efficacy is more important than the color rendering, then one would choose the trichromatic phosphors (these phosphors produce three narrow bands of light in the primary colors). If a warm white color is desired, then a powder is employed that emits a spectrum with emphasis on the red part, while for cold white, the emphasis shifts to blue.

The first effort to produce visible photons using a phosphor took place in 1934 by GE and Osram research laboratories, but the major breakthrough occurred with the development of the calcium halophosphate phosphor, $Ca_5(PO_4)_3(Cl,F):Sb^{3+},Mn^{2+}$, in 1942. This type of phosphor combines two complementary color emission bands into one phosphor.

The spectral output of a commercial fluorescent lamp with a phosphor coated on the inside wall can be seen in Figure 4.7. The spectrum shows mercury's emission lines and the phosphor's continuum in the red area.

The efficiency of phosphors used today in fluorescent lamps is quite high, exhibiting quantum efficiencies of more than 0.85. The term *quantum efficiency* describes the ratio of visible photons emitted divided by the

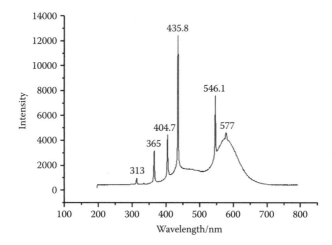

FIGURE 4.7 Emission spectrum of a low-pressure discharge lamp with phosphor coated on the inside wall.

number of UV quanta absorbed. Nonradiative losses can occur when the emitting state relaxes to the ground state via a crossover of the potential energy curves and is more likely to happen at higher temperatures.

Of course, there are applications that require UV light, such as sterilization of rooms and equipment. In this case, the resonance line is sufficient, and so the lamp is used without a phosphor (the bulb in this case should be made of quartz, as soft glass blocks radiation with wavelengths below 300 nm due to the sodium carbonate it contains) and tanning lamps when used powders emitting the UV-A (310–380 nm). Another UV application is tanning and, in this case, the phosphors employed convert the resonance lines of mercury into UV-A radiation (310–380 nm).

When the collisions between the free electrons and mercury atoms are elastic, we have momentum transfers and changes in velocity (changes in the direction of the moving particles). Such collisions, where only kinetic energy is exchanged, result in increases of the temperature of the atomic gas (mercury vapor).

If the colliding electron has an energy equal to or higher than the ionization energy of the atom, then ionization occurs, which is necessary for the maintenance of the discharge.

In the third case, the collision between the electron and the atom results in the excitation of the atom, and one of the atom's valence electrons quantum leaps to a higher orbit or, in other words, to a higher energy state. In mercury's case, the electrons must have energy of at least 4.89 eV in order

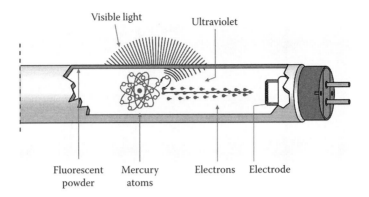

FIGURE 4.8 Anatomy of a mercury vapor discharge tube known as the fluorescent lamp.

to cause excitation of the mercury atom and 10.43 eV in order to cause ionization. The average energy of an electron in a low-pressure discharge tube is 1.1 eV, which corresponds to the mean temperature of the electron cloud (around 11,000 K), but a percentage of them (the high end tail of the distribution curve) have enough energy to cause inelastic collisions (excitation or ionization).

Figure 4.8 shows the anatomy of a low-pressure fluorescent lamp and the steps involved in the emission of light. Electrons leave the electrodes, travel through space, collide with mercury atoms, radiation is emitted by the mercury atoms, and finally the wall phosphor converts some of that radiation into visible light.

4.2.1.2 The Electrodes

Electrodes play an important role in starting the discharge. On initially applying a voltage across the tube, there is practically no ionization and the gas behaves as an insulator. Once a few ions or electrons are present, a sufficiently high voltage accelerates them to provide more carriers by electron impact ionization, and breakdown is achieved. Suitable cathodes supply electrons at a very early stage in the breakdown process by field emission, photoelectric emission, or thermionic emission, greatly reducing the excess voltage needed to strike the discharge. If the electrodes are preheated or if some arrangement, such as an adjacent auxiliary electrode, is provided, the voltage required is further reduced.

The electrodes must be good emitters of electrons and must be capable of operating at high temperatures. They should also have a low evaporation rate and high sputtering resistance. This is why tungsten is usually

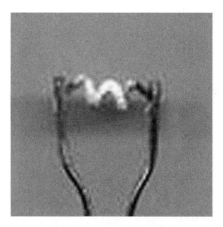

FIGURE 4.9   Fluorescent lamp electrode (tungsten with coating).

used. However, in order to extend the lamp life, the operating temperature should not be very high, so other electron-emissive materials (coated on tungsten) have to be used. The materials most often used are the oxides of calcium, strontium, and barium, which is the coating of the lamp electrodes used in this study (Figures 4.9–4.11). The operation of the discharges under AC also ensures a longer life span since both electrodes act as cathodes, sharing the workload.

As the triple oxide is very reactive with air, a mixture of the carbonates is coated on the coil, which is inert. The lamps can later be processed under vacuum by heating in order to decompose the carbonates into the oxide form as it is the latter that has a low work function and good

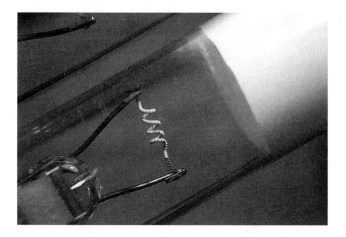

FIGURE 4.10   Fluorescent lamp electrode.

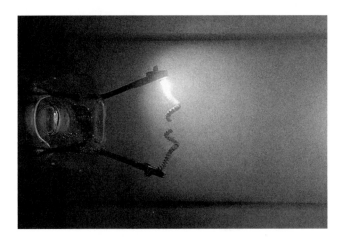

FIGURE 4.11   Fluorescent lamp electrode.

emissive properties. The maximum gas evolutions for the three carbonate compounds occur at the following temperatures:

$$CaCO_3 = 1062 \text{ K}$$

$$SrCO_3 = 1166 \text{ K}$$

$$BaCO_3 = 1206 \text{ K}$$

Table 4.1 lists the work function values of various compounds used in the construction of discharge lamp electrodes. The work function is the energy needed for an electron to be released from the compound.

The dimensions of the electrode and, more specifically of the tungsten filament, depend on the application of the lamp and its power.

4.2.1.3 The Ballast

An operating gas discharge is a nonlinear circuit element that cannot be described by Ohm's law ($V = IR$) because the conductivity increases

TABLE 4.1   Work Function Values of Electrode Materials

Cathode Material	Work Function/eV
Tungsten/W	4.54
Barium/Ba	2.5
Barium oxide/BaO	1
Strontium oxide/SrO	1.45
Calcium oxide/CaO	1.75

with current. The discharge is therefore said to exhibit negative differential voltage–current characteristics, except in some special cases such as weakly ionized plasmas. A current-limiting device, called a *ballast*, is therefore necessary for the operation of discharge lamps. In a discharge tube, the low-pressure gas between the electrodes acts as an insulator, and it is only with the application of high voltage that the gas begins to ionize and becomes conductive. As the current intensity increases, the plasma (ionized gas) resistance decreases due to the larger number of ions, so we need a ballast to control the current to a desired value before it tends to infinity and the tube is destroyed.

The function of a ballast is also to apply sufficient voltage to ignite the lamp as well as the regulation of the lamp current and, in the case of AC operation, to relight the lamp each half-cycle.

A simple series-connected resistor can be sometimes used as a ballast, but the power losses (I^2R) result in a low overall efficiency. The use of a resistor on an alternating supply leads to reignition delays, causing near-zero current periods at the start of each half-cycle. The advantage offered by AC operation, however, is that an inductive or capacitive impedance can be used to provide current limitation, resulting in a significant energy loss reduction.

Standard fluorescent magnetic ballasts use a combination of inductive and capacitive networks for current control, which essentially comprises an aluminum coil wrapped around an iron core. This combination reduces power losses even more, by reducing the phase difference between the voltage and current waveforms.

The energy-efficient magnetic ballast is an improved version of the standard one employing copper wire instead of aluminum wrapped around larger iron cores. The efficiency is improved due to copper's lower resistance and less heat generated by the larger iron core.

Modern electronic ballasts eliminate the large, heavy iron ballast and replace it with an integrated inverter/switcher. They operate in a similar manner as iron ballasts but at a much higher frequency (>20 kHz instead of 50 Hz, which magnetic ballasts operate at). Power is provided to the ballast at mains frequency and is converted to a few tens of kilohertz.

Current limiting is achieved by a very small inductor with high impedance at these high frequencies. The higher frequency results in more efficient transfer of input power to the lamp, less energy dissipation, and elimination of light decay during each cycle. For economic reasons, a

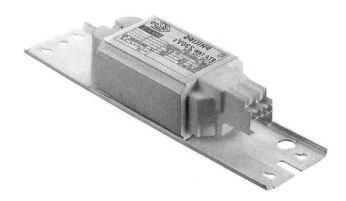

FIGURE 4.12 Electronic ballast for a fluorescent lamp.

combination of a standard and an electronic ballast is manufactured, usually referred to as a *hybrid ballast*, offering good energy efficiency but reduced light output and lamp life.

The trends of the market show that electronic ballasts, such that shown in Figure 4.12, will dominate, and soon all discharge lamps of any pressure will employ them.

Along with controlling the intensity of current to a desired value, the ballast/gear also manages the starting of the lamp, which is achieved in two ways. In one method, the electrodes are heated, thereby reducing the initial voltage needed for ignition, but this way the lamp consumes more energy. In the other method, the electrodes are not heated, thus saving energy, but the applied voltage needs to be higher; so the electrodes are subject to greater wear, thus reducing the average life of the lamp. The choice of starter depends on the use of the lamp, so, when multiple switches are involved, the heated electrodes are preferred. Finally, the electronic ballast offers the option of dimming all the way down to a few percent of the maximum brightness.

The tubular fluorescent lamps, such as those in Figure 4.13, come in various lengths depending on the wattage of the lamp, and they are known by a code (such as TL) followed by a number indicating the diameter of the tube in eighths of an inch. The most widespread are the TL8 (1 in. diameter), but they are losing popularity to the TL 5. Figures 4.14–4.22 show representative commercial fluorescent lamps and their typical emission spectra, while Tables 4.2–4.6 show some of their typical lamp characteristics. Throughout this chapter, there will be emission spectra and characteristics tables of representative discharge lamps.

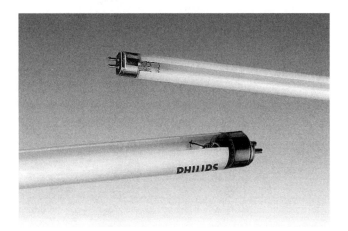

FIGURE 4.13 Tubular fluorescent lamps.

Another broad category of lamps are the compact fluorescent lamps, CFL, which are increasingly used as replacements of incandescent bulbs due to their energy savings. The placement of the electronic ballast at the base of the lamp makes them even more practical.

In general, fluorescent lamps are much more economical than incandescent lamps with a luminous efficacy up to four times higher than the most efficient halogen bulb and a much longer average life. For this reason, they are the preferred choice for applications where color rendering has

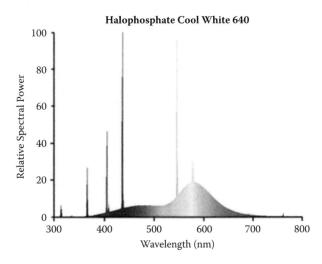

FIGURE 4.14 Emission spectrum example of a fluorescent lamp with a phosphor.

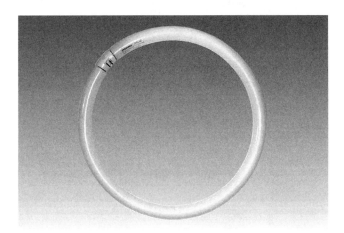

FIGURE 4.15   Circular fluorescent lamp (Sylvania Circline ring-shaped 60 W).

to be good but not great—high brightness is required and replacement is not easy (the induction lamp, in particular, is useful for this latter reason). However, it must be noted that there are fluorescent lamps in the market that show near-perfect color rendering properties with the appropriate phosphors but at the expense of efficacy. Fluorescent lamps have dominated the lighting market and are mainly used for general lighting in offices, living spaces, and other large areas such as warehouses.

From the fluorescent lamp, which is a low-pressure mercury vapor discharge tube, we can move on to another category of lamps that operate on the same principle of low-pressure electrical discharges but where the active media are different, thus giving us different emission spectra [3,4].

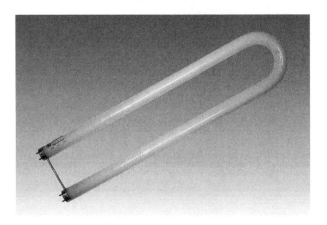

FIGURE 4.16   U-shaped fluorescent lamp (GE U-shaped 40 W).

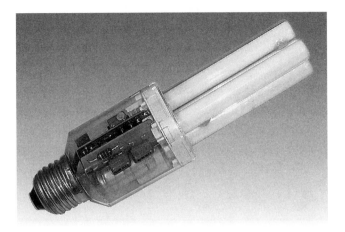

FIGURE 4.17 Compact fluorescent lamp with integrated electronic gear/ballast (Philips PLCE 7 W).

4.2.2 Low-Pressure Noble Gas Discharges

If we replace mercury with a range of gases such as the noble gases, then we have discharges with instant starting because the active medium is already in the gas phase and is not affected by ambient temperature or the temperature of the lamp. Although the noble gases (helium, neon, argon, krypton, and xenon) emit in the visible range, their resonance lines (see Table 4.7) that constitute most of the emitted radiation are in the ultraviolet and even below 200 nm; so appropriate fluorescent powders to convert

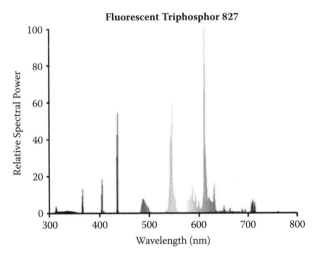

FIGURE 4.18 Emission spectrum of a compact fluorescent lamp.

FIGURE 4.19   Compact fluorescent lamp (Philips "Tornado" Helical-shape 23 W).

this radiation into visible light would be needed. However, the conversion of radiation from UV-C to visible would mean large Stokes losses, and this is why such lamps are used mainly for decoration and signs and not for general lighting, as shown in Figure 4.23.

Of all the noble gases, neon is used most frequently as an active medium in low-pressure discharge tubes. The reason for this is that neon emits, besides its resonance lines and with a lower efficiency, a large number of visible emission lines in the red region of the spectrum, so neon lamps without employing any phosphors have found

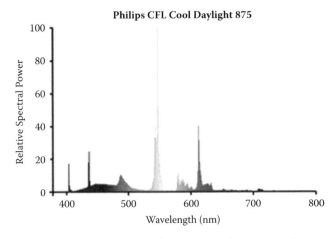

FIGURE 4.20   Emission spectrum of a compact fluorescent lamp.

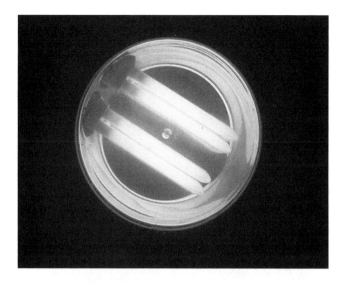

FIGURE 4.21   Emission of light from compact fluorescent lamps.

applications in advertising signs. Figures 4.24 and 4.25 show neon's red emissions.

Neon is also used in indicator lamps of electronic devices, while the rest of the gases are used in more specific applications such as discharge tubes for lasers. Xenon under pulsed conditions (microsecond scale and

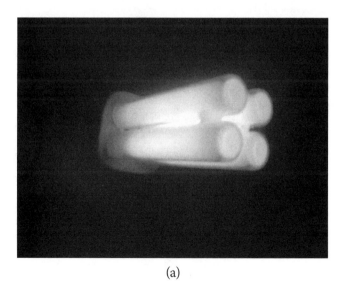

(a)

FIGURE 4.22   Variety of compact fluorescent lamps.

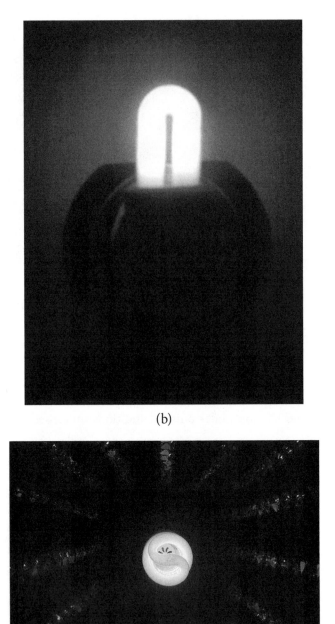

(b)

(c)

FIGURE 4.22   (*Continued*)

TABLE 4.2   Indicative Values and Information about the
General Characteristics of the Tubular Fluorescent Lamp

Gas Fill	Ar	Hg
Luminous flux	5,000 lm (100 h)	
Efficacy	100 lm/W (100 h)	
Color temperature and CRI	CCT: Wide range	CRI: Ra=>80
Average lifetime	20,000 h	

TABLE 4.3   Table with Indicative Values and Information about
the General Characteristics of the Circular Fluorescent Lamp

Gas Fill	Ar	Hg
Luminous flux	3,400 lm (100 h)	
Efficacy	56 lm/W (100 h)	
Color temperature and CRI	CCT: 3,000 K	CRI: Ra = 53
Color coordinates	CCx: 0.436	CCy: 0.406
Average lifetime	11,000 h	

TABLE 4.4   Table with Indicative Values and Information about
the General Characteristics of the U-Shaped Fluorescent Lamp

Gas Fill	Ar	Hg
Luminous flux	2,875 lm (100 h)	
Efficacy	72 lm/W (100 h)	
Color temperature and CRI	CCT: 3,450 K	CRI: Ra = 54
Average lifetime	12,000 h	

TABLE 4.5   Table with Indicative Values and Information about
the General Characteristics of a Compact Fluorescent Lamp

Gas Fill	Ar–Kr	Hg
Luminous flux	400 lm (100 h)	
Efficacy	57 lm/W (100 h)	
Color temperature and CRI	CCT: 2700 K	CRI: Ra = 82
Color coordinates	CCx: 0.473	CCx: 0.420
Average lifetime	8000 h	

TABLE 4.6   Indicative Values and Information about the
General Characteristics of a Compact Fluorescent Lamp

Gas Fill	Ar	Hg
Luminous flux	1450 lm (100 h)	
Efficacy	63 lm/W (100 h)	
Color temperature and CRI	CCT: 7350 K	CRI: Ra = 83
Color coordinates	CCx: 0.303	CCx: 0.310
Average lifetime	6000 h	

TABLE 4.7   Table with Ionization Energies and Resonance Line Wavelengths
of Various Active Media

Element	Metastable Levels (eV)	Ionization Energy (eV)	Resonance Emission Lines/nm
Neon	16.53, 16.62	21.56	73.6, 74.4
Argon	11.49, 11.66	15.76	104.8, 106.7
Krypton	9.86, 10.51	13.99	116.5, 123.6
Xenon	8.28, 9.4	12.13	129.5, 146.9
Mercury	4.64, 5.44	10.43	184.9, 253.7
Sodium	—	5.14	589.0, 589.6

FIGURE 4.23 **(See color insert following page 20.)** Low-pressure gas
discharge lamps for decoration and signs.

FIGURE 4.24 **(See color insert following page 20.)** The neon discharge lamp has dominated the advertising signs industry for decades.

frequency of several hundred hertz) and low pressure (several tens of torr) generates short bursts of high-intensity white light and finds applications in photography (Figure 4.26) and spectrum (Figure 4.27).

A similar technology is that of the dielectric barrier discharge (DBD), which does not utilize the resonance lines of xenon atoms (129 and 147 nm) for the excitation of the fluorescent powder but the emission of xenon dimers (172 nm) generated under the conditions of operation. The visible light emitted by the powder in combination with the visible light emitted by the active medium gives us a light source of low efficacy (30 lm/W) compared to existing fluorescent lamps but of long average life (over 100,000 h), and it offers the advantage of avoiding environmentally harmful materials such as mercury (Table 4.8). This technology (DBD Xenon) has already been channeled into the market in the form of a flat light source (surface area of 21 in.) for general lighting or liquid crystal display backlight (Figures 4.28 and 4.29).

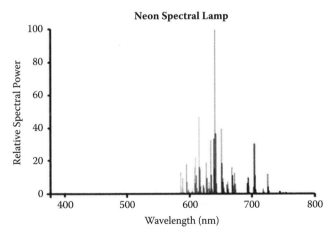

FIGURE 4.25 Low-pressure neon discharge lamp emission spectrum.

FIGURE 4.26   Photographic xenon flash during light emission.

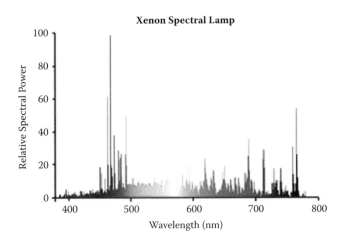

FIGURE 4.27   Low-pressure xenon emission spectrum.

TABLE 4.8   Indicative Values and Information about the General Characteristics of the Low-Pressure Xenon Lamp

Gas Fill	Xe_2	
Luminous flux	1,850 lm (100 h)	
Efficacy	27 lm/W (100 h)	
Color temperature and CRI	CCT: 7,400 K	CRI: Ra = 86
Color coordinates	CCx: 0.304	CCy: 0.295
Average lifetime	100,000 h	

FIGURE 4.28 Dielectric barrier discharge lamp (DBD) with xenon at low pressure for general lighting (Osram Planon 68 W).

Based on the emissions of the xenon dimer molecules under pulsed conditions of operation and the use of a fluorescent powder to convert ultraviolet light into visible, an electroded discharge tube has been developed for general lighting, although the efficacy is still low compared to low-pressure mercury vapor discharge lamps. An interesting development of this technology is the use of multiple electrodes [5] in order to maintain plasma diffusion and homogeneity and prevent the plasma inside the tube from constricting.

Dielectric barrier discharge (DBD) excimer or exciplex lamps are known as efficient UV and VUV sources with narrow spectra. Another example of DBD development is the Xe-Cl_2 DBD lamp, which produces more than 90% of its radiated power at 308 nm, which corresponds to the emission of Xe*Cl exciplex molecule.

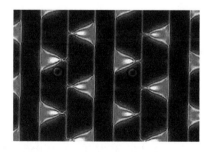

FIGURE 4.29 Plasma formation in xenon DBD.

There is ongoing research and development of xenon discharge lamps under pulsed operation for general lighting due to the pressure to ban environmentally harmful materials such as mercury, and several references in scientific journals show that their luminous efficacy has increased.

4.2.3 Low-Pressure Sodium Vapor Discharges

The low-pressure discharge lamp with the highest luminous efficacy—up to 200 lm/W—is that of sodium vapor. This high efficiency is due to the fact that the resonance line of sodium (Figures 4.30 and 4.31), produced after excitations of sodium atoms by electron collisions, is at 589 nm, that is, in the visible spectrum and does not need conversion by a fluorescent powder.

The lamp is made of hard boron glass to withstand the corrosiveness of sodium and also contains a neon–argon mixture that acts as the buffer gas. The partial pressure of sodium is much lower than that of mercury, so, for the proper operation of the lamp, higher temperatures are required than that of the fluorescent lamp. To increase the temperature, a second outer tube encloses the inner one in a vacuum, and a special coating is applied on the inner walls of the outer tube that allows the escape of visible light but reflects back the infrared radiation (heat).

One of the reasons the low-pressure sodium lamp has a high efficacy is because sodium's resonance line lies around the peak of the human eye's sensitivity curve (589 nm), making it at the same time inappropriate for

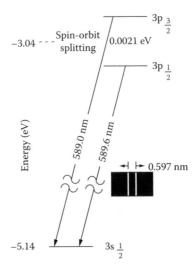

FIGURE 4.30 Simplified energy level diagram of sodium showing the resonant states.

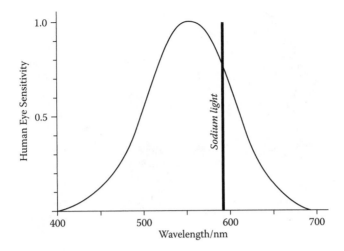

FIGURE 4.31 Sodium's resonance line lies around the peak of the human eye's sensitivity curve (589 nm).

domestic lighting due to its poor color rendering properties. The CRI is a relative measure of the shift in surface color of an object when lit by a particular lamp, compared with how the object would appear under a reference light source of similar color temperature.

The low-pressure sodium lamp has a CRI of zero (and low color temperature of around 1700 K), so it is mainly used in applications where high brightness is required but color reproduction is not so important, such as street lighting, safety lighting, lighting of large outdoor spaces (parking spaces), etc. Figures 4.32 and 4.33 show examples of street lighting employing low-pressure sodium vapor discharge lamps. Table 4.9 summarizes the characteristics of the low pressure sodium lamp.

A good candidate to replace existing active media in discharge lamps must be an efficient emitter in the range 380–580 nm. The reason for selecting this particular range is the need to eliminate any Stoke losses that would be associated with photons the human eye cannot detect and would need conversion using a phosphor. On the short-wavelength side, the eye sensitivity, and therefore the visible range, starts at around 380 nm. On the long-wavelength side, perhaps there is no need to include wavelengths beyond 580 nm as sodium is already a strong emitter just above that limit, the sensitivity of the eye drops, and, in any case, the photons on that part of the spectrum cannot yet be converted by using phosphor. On the other hand, strong emissions in the blue region could be partly converted in order to cover larger regions of the visible spectrum.

FIGURE 4.32 Low-pressure sodium vapor discharge lamps for street and road lighting.

The average lifetime of these lamps reaches nearly 20,000 h. Until the temperature reaches the required levels where the necessary amount of sodium vaporizes, light is emitted by the excited buffer gas neon atoms, which explains the red color of the lamp light during its first minutes of operation after each start. As sodium vaporizes and more of its atoms participate in the electrical discharge process, the lamp gets its characteristic yellow color due to radiation at 589 nm.

FIGURE 4.33 **(See color insert following page 20.)** Low-pressure sodium vapor discharge lamps for street and road lighting.

FIGURE 4.34   Low-pressure sodium vapor discharge lamp (Philips SOX 35 W).

4.3 PRESSURE EFFECT

Figures 4.34–4.36 show a representative commercial low-pressure sodium vapor lamp along with a spectrum, while Table 4.9 shows the characteristics of such lamps.

4.3.1 Thermal Equilibrium

All plasmas have a number of features in common, but one way to begin classification is by quantifying the density of the charge carriers and the thermal energy of the electrons in particular. One of the plasma parameters that has a pronounced effect on both of these quantities is the pressure of the gas.

FIGURE 4.35   Electrode of low-pressure sodium vapor discharge lamp.

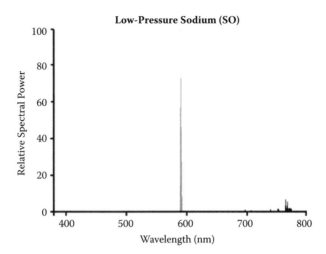

FIGURE 4.36 Emission spectrum of a low-pressure sodium vapor discharge lamp.

One of the principal characteristics of low-pressure discharges is the absence of thermal equilibrium in the plasma. While a low-pressure discharge is collision-dominated plasma, the collision rate is generally not high enough for the electrons to equilibrate with the heavy atoms. In all cases, the electron temperature is much greater than the kinetic temperature of the atoms, ions, and molecules. An average value for the electron temperature in low-pressure discharges is 11,000 K, which corresponds to 1.0 eV, but a fraction of the electrons will possess many times this energy, enabling them to cause excitation and ionization.

The atomic gas, on the other hand, will only be at a temperature of a few hundred degrees, but the energy transfer between the electron and atomic gas becomes more significant at higher pressures, at which the collision rate increases dramatically. A point is then reached where the temperatures of the two gases reach an average value, and local thermal equilibrium is then established (Figure 4.37).

4.3.2 Line Broadening

The pressure increase also affects the width of lines. In a low-pressure discharge, the width of the atomic lines is Doppler dominated. That means that the lines have a finite width due to their thermal motion but, at pressure of more than a few atmospheres, the lines are significantly broadened due to the perturbation of the energy level by other nearby atoms. This perturbation is even more pronounced when the interaction is between identical atoms.

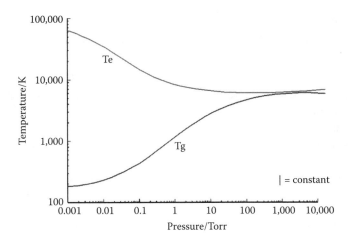

FIGURE 4.37 Gas and electron temperatures with respect to gas fill pressure. At higher pressures, conditions of local thermal equilibrium are achieved.

An example of this is the high-pressure sodium lamp, where the color rendering properties of the lamp are greatly improved due to the occurrence of resonant broadening, as we shall see later.

4.4 HIGH-PRESSURE DISCHARGE LAMPS

When one requires high brightness, relatively high efficiency, and good color rendering, high-pressure discharge lamps can be used. The lamps operate using the same principle of electric discharge in a gas or vapor that results in excitation of the active medium atoms and the subsequent emission of radiation. As the name of this group of lamps indicates, the difference is that the gas or vapor pressure is much higher (several atmospheres).

In a high-pressure discharge tube, the wattage per discharge centimeter (power density) is much higher than that of a low-pressure discharge tube, so the number of collisions between electrons and atoms is also much

TABLE 4.9   Table with Indicative Values and Information about the General Characteristics of the Low-Pressure Sodium Vapor Lamp

Gas Fill	Inner: 1% Ar in Ne	Na	Outer: Vacuum
Luminous flux	4,800 lm (100 h)	4,560 lm (end of lifetime)
Efficacy	126 lm/W (100 h)	120 lm/W (end of lifetime)
Color temperature and CRI	CCT: 1,700 K	CRI: Ra = 0
Color coordinates	CCx: 0.574	CCy: 0.425
Average lifetime	20,000 h	

TABLE 4.10   General Comparisons of Low- and High-Pressure Discharge Lamps

Low-Pressure Discharge Tube	High-Pressure Discharge Tube
Electron temperature = 11,000 K	Electron temperature = 5,000 K
Gas temperature = 350 K	Gas temperature = 5,000 K
Mainly emissions of the resonance lines	Emission lines from higher-energy states
Usually long tube	Usually short tube
Low brightness	High brightness
Pressure lower than atmospheric	Pressure higher than atmospheric
Power less than 1 kW	Power of up to several kilowatts
Tube made usually of soft glass	Tube usually made of quartz or polycrystalline alumina (PCA)

higher. The higher number of collisions, the higher the gas temperature, due to transfer of kinetic energy and the resulting increased gas pressure. The electrodes emit electrons due to their high temperature (thermionic emission) without the aid of an electric field, and the emissions of the active medium atoms are not limited to the resonance lines. Table 4.10 summarizes the differences between low- and high-pressure discharges.

The high-pressure discharge lamp consumes more energy than a low-pressure one and is characterized by a far greater efficiency than an incandescent lamp of equal power. This category of lamps is known as HID (high-intensity discharge) lamps. Similar to the low-pressure discharge lamps, they require a ballast to keep the intensity of the current to a desired value. There are many types of gear, and their choice depends on the power consumption of the lamp and the external operating conditions.

A high-pressure discharge starts with pulses of high voltage applied between the two electrodes while, in some tubes, a third electrode is placed adjacent to one of the others so that a smaller discharge causes ionization and makes the gas or vapor conductive. The buffer gas also plays a role in creating a discharge and starting the lamp. Because of their high brightness and efficacy, these lamps are employed for lighting large spaces such as warehouses, stadia, public places, parking lots, and roads. Another application where we encounter such high-pressure discharge lamps is automotive lighting, where they replace conventional incandescent lamps as front lights.

4.4.1 High-Pressure Mercury Vapor Discharges

Mercury is used as the active medium, not only in fluorescent lamps but also in high-pressure discharges. The glass of the bulb is quartz, which is more resistant to high temperatures and pressures, and it is also enclosed into another larger glass body that is under vacuum. Due to the higher

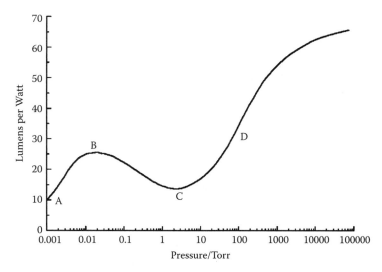

FIGURE 4.38   The luminous efficiency of a gas discharge varies with the Hg vapor fill pressure.

number of electron impacts, the atoms of mercury are excited to high-energy states, leading to emission spectra with visible lines of higher intensity than those found in the spectrum of a fluorescent lamp.

Figure 4.38 shows the dependence of efficacy on mercury vapor pressure. At point A, the efficacy is very low since the conditions there favor the production of the ultraviolet resonant lines. With increasing pressure, however, absorption of the resonant lines causes an increase in the population of higher-energy levels. Since these levels include the visible mercury lines, the increase in their population results in an improvement of the efficacy of the discharge. The decrease in the efficacy beyond point B is due to the fact that elastic collisions between electrons and mercury atoms increase with increasing pressure, so more energy is lost due to elastic collisions and efficacy decreases; at the same time, the gas temperature rises, resulting in transfer of the energy input to the tube walls due to thermal conduction. Although the gas temperature increases with increasing pressure, the electron temperature decreases as a result of the larger number of collisions.

When the pressure exceeds point C, the gas temperature has become so high that thermal excitation of mercury atoms is possible, so the luminous efficiency of the discharge increases again. So, the "low-pressure" and "high-pressure" discharges differ considerably from each other in their fundamental mechanism, because the former consists mostly of resonant transitions, while the latter has transitions between the upper energy levels. Perhaps, a better distinction between these two kinds of discharges would be "nonthermal" and "thermal."

Once again we deal with a line spectrum and not a continuous one, so the final result is sometimes formed using a fluorescent powder that converts any UV radiation emitted into visible light. The phosphor is coated on the inside wall of the outer tube that serves as a thermal insulator. In general, the spectrum is deficient in emissions in the red part, and the characteristics of such a lamp are defined by the choice of phosphor. Some indicative values for the lamp characteristics are luminous efficacy of 60 lm/W, CRI of 15–55, color temperature at around 3,000–4,000 K, and average lifetime of about 15,000 h. Figures 4.39–4.43 show commercial high-pressure mercury vapor lamps and typical emission spectra, while Tables 4.11 and 4.12 list some general characteristics.

A variation of the high-pressure mercury vapor lamp is the super high-pressure mercury lamp (200 atm), which does not employ a secondary outer tube and is used in projectors for image and video displays (Figure 4.44, spectrum in Figure 4.45, and characteristics in Table 4.13). In this case, the mercury emission lines are strongly broadened due to the high pressure (pressure broadening) and cover the entire visible spectrum, resulting in a high-intensity white light and obviating the need for fluorescent powders.

Yet another variation is the hybrid lamp, which combines the mercury vapor discharge and incandescence (Figure 4.46, spectrum in Figure 4.47, and characteristics in Table 4.14). The lamp emits a warm white light with

FIGURE 4.39 High-pressure mercury vapor discharge lamp (Iwasaki H2000B Clear Hg 2000 W).

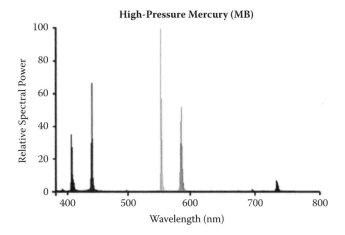

FIGURE 4.40 An emission spectrum of a high-pressure mercury vapor discharge lamp.

low luminous efficacy and poor color rendering. The incandescent filament, which contributes to the light emissions (in the red part), also plays the role of current intensity regulator as a resistance; that is, the lamp is self-ballasted.

4.4.2 High-Pressure Sodium Vapor Discharges

High-pressure sodium vapor discharge tubes in fact employ a mix of sodium and mercury, resulting in emission of radiation with greater coverage of the visible spectrum. Another difference from the high-pressure mercury

FIGURE 4.41 High-pressure mercury vapor discharge lamp (Philips HPL + YAG + Yttrium Vanadate coating 80 W).

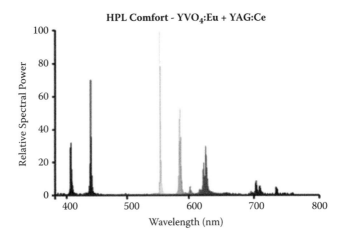

FIGURE 4.42 An emission spectrum of a high-pressure mercury vapor discharge lamp with phosphor.

FIGURE 4.43 Electrode of a high-pressure mercury vapor discharge lamp.

TABLE 4.11 Table with Indicative Values and Information about the General Characteristics of the High-Pressure Mercury Vapor Lamp

Gas Fill	Inner: Ar | Hg	Outer: N_2
Luminous flux	120,000 lm (100 h)	
Efficacy	60 lm/W (100 h)	
Color temperature and CRI	CCT: 5,700 K	CRI: Ra = 15
Average lifetime	10,000 h	

TABLE 4.12   Table with Indicative Values and Information about the General Characteristics of a High-Pressure Mercury Vapor Lamp

Gas Fill	Inner: Ar | Hg	Outer: N_2
Luminous flux	4,000 lm (100 h)	
Efficacy	50 lm/W (100 h)	
Color temperature and CRI	CCT: 3,500 K	CRI: Ra = 57
Color coordinates	CCx: 0.40	CCy: 0.38
Average lifetime	28,000 h	

FIGURE 4.44   Super high-pressure mercury vapor discharge lamp (Philips UHP LCD video projector 120 W).

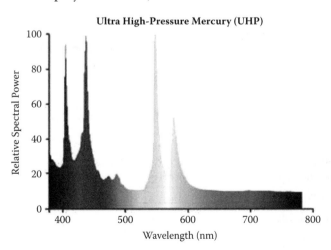

FIGURE 4.45   The emission spectrum of a super high-pressure mercury vapor discharge lamp.

FIGURE 4.46 Hybrid high-pressure mercury lamp (Philips ML Blended Self-ballasted 160 W).

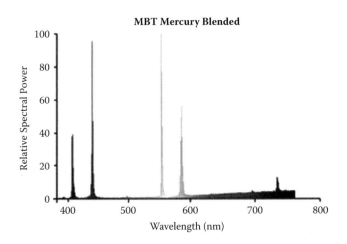

FIGURE 4.47 Emission spectrum of a hybrid mercury lamp.

TABLE 4.13 Table with Indicative Values and Information about the General Characteristics of the Super High-Pressure Mercury Vapor Lamp

Gas Fill	Inner: Ar–Br	Hg	
Luminous flux	7000 lm (100 h)	
Efficacy	58 lm/W (100 h)	
Color temperature and CRI	CCT: 7600 K	CRI: Ra = 57
Color coordinates	CCx: 0.298	CCy: 0.311
Average lifetime	6000 h	

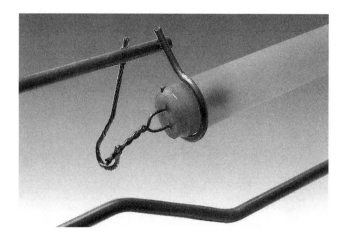

FIGURE 4.48   Polycrystalline alumina burner (PCA).

lamps is the material used for the construction of the inner bulb that contains the metallic active media. Quartz used to manufacture high-pressure mercury lamps and boron glass used to manufacture low-pressure sodium lamps cannot withstand the corrosivity of sodium under high temperatures and pressures, so a crystalline material of aluminum oxide (polycrystalline alumina, or PCA) is what is used. Lamps made from this material are also known as *ceramic burners*, and Figure 4.48 depicts one.

Due to their continuous emission spectrum, the high-pressure sodium vapor lamps have a good CRI of up to 85, and the luminous efficacy is quite high and exceeds 100 lm/W (taking into account the sensitivity curve of the eye, the efficacy reaches 150 lm/W). The color temperature is around 2,000–2,500 K, which is a warm white, and the average life of the bulb can be up to 30,000 h. Their applications are the same as the high-pressure mercury vapor lamps, except that high-pressure sodium lamps offer better color rendering. Figures 4.49, 4.51, 4.53, and 4.55 show commercial high-pressure sodium vapor discharge lamps, while their respective typical spectra are shown in Figures 4.50, 4.52, 4.54, and 4.56. Tables 4.14–4.18 list some of their general lamp characteristics. Table 4.19 summarizes the different high-pressure sodium lamp designs.

4.4.3 Metal Halide Discharge Lamps

A further development of the high-pressure mercury or sodium vapor discharge lamps is the addition of more active media in order to tailor the emission spectrum and have control over the CRI or the color temperature. The addition of other pure metals known for their number of emission lines in the

FIGURE 4.49 High-pressure sodium vapor discharge lamp (reflux with high special side reflector 250 W).

visible spectrum would not have the desired effect because of their very low partial pressures, even at temperatures encountered in high-pressure tubes.

A solution to overcome the problem of low vapor pressures of the pure metals is the addition of their halide compounds that are easier to vaporize. Once in the gas phase, the compounds are dissociated due to electron collisions, and the freed metallic atoms are excited. The excitation is followed by relaxation of the atoms, and the extra energy is released in the form of

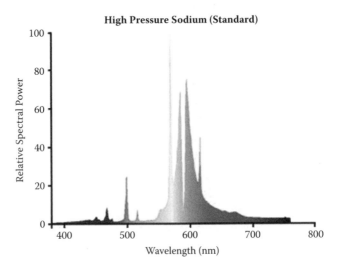

FIGURE 4.50 Emission spectrum of a high-pressure sodium vapor discharge lamp.

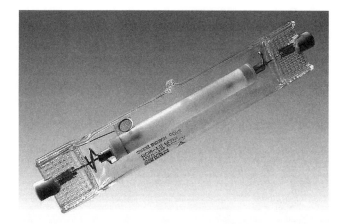

FIGURE 4.51   High-pressure sodium vapor discharge lamp (SON-TS Vialox Super Vacuum Linear 150 W).

emitted electromagnetic waves with a frequency depending on the metal of choice and its signature energy level structure.

$$MX_n \leftrightarrow M + nX$$

$$M + e^- \rightarrow M^*$$

$$\rightarrow M^+ + 2\,e^-$$

$$M^* \rightarrow M + h\nu$$

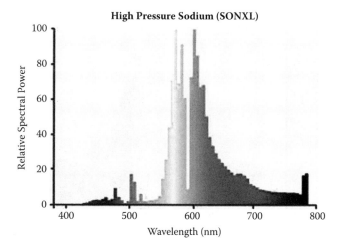

FIGURE 4.52   Emission spectrum of a high-pressure sodium vapor discharge lamp.

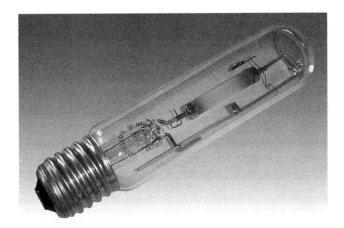

FIGURE 4.53 High-pressure sodium vapor discharge lamp (Philips SON city beautification 150 W).

where *M* is the metal, *X* is the halogen, *n* is the stoichiometric number, *h* is Planck's constant, and *v* is the frequency of the emitted wave.

The combination of emitted radiation from various metallic elements allows the development of lamps with a variety of features. These lamps that are called *metal halide lamps* have a CRI of up to 90, and their color temperatures start at 3,000 K and reach 20,000 K. The luminous efficacy is generally high with over 110 lm/W for some lamps, and the average

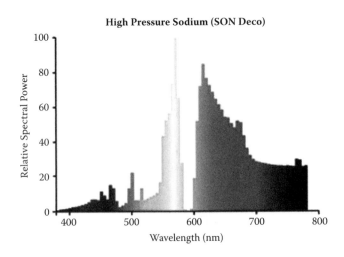

FIGURE 4.54 Emission spectrum of a high-pressure sodium vapor discharge lamp.

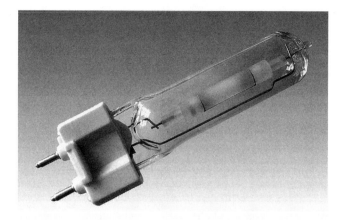

FIGURE 4.55   High-pressure sodium vapor discharge lamp (Philips SDW-TG Mini White 100 W).

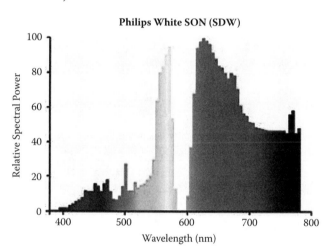

FIGURE 4.56   Emission spectrum of a high-pressure sodium vapor discharge lamp.

TABLE 4.14   Table with Indicative Values and Information about the General Characteristics of the Hybrid Mercury Lamp

Gas Fill	Inner: Ar | Hg	Outer: N_2
Luminous flux	3000 lm (100 h)	
Efficacy	18 lm/W (100 h)	
Color temperature and CRI	CCT: 3800 K	CRI: Ra = 60
Color coordinates	CCx: 0.349	CCy: 0.380
Average lifetime	6000 h	

TABLE 4.15   Table with Indicative Values and Information about the
General Characteristics of a High-Pressure Sodium Vapor Lamp

Gas Fill	Inner: Xe | Na, Hg	Outer: Vacuum
Luminous flux	26,000 lm (100 h)	
Efficacy	104 lm/W (100 h)	
Color temperature and CRI	CCT: 1,900 K	CRI: Ra = 25
Color coordinates	CCx: 0.542	CCy: 0.415
Average lifetime	24,000 h	

TABLE 4.16   Table with Indicative Values and Information about the
General Characteristics of a High-Pressure Sodium Vapor Lamp

Gas Fill	Inner: Xe | Na, Hg	Outer: Vacuum
Luminous flux	15,000 lm (100 h)	
Efficacy	100 lm/W (100 h)	
Color temperature and CRI	CCT: 2,000 K	CRI: Ra = 25
Color coordinates	CCx: 0.530	CCy: 0.430
Average lifetime	12,000 h	

TABLE 4.17   Table with Indicative Values and Information about the
General Characteristics of a High-Pressure Sodium Vapor Lamp

Gas Fill	Inner: Xe | Na, Hg	Outer: Vacuum
Luminous flux	7000 lm (100 h)	
Efficacy	48 lm/W (100 h)	
Color temperature and CRI	CCT: 2500 K	CRI: Ra => 80
Average lifetime	8000 h	

TABLE 4.18   Table with Indicative Values and Information about the
General Characteristics of a High-Pressure Sodium Vapor Lamp

Gas Fill	Inner: Na, Hg/Xe	Outer: Vacuum
Luminous flux	4,800 lm (100 h)	
Efficacy	48 lm/W (100 h)	
Color temperature and CRI	CCT: 2,500 K	CRI: Ra = 83
Color coordinates	CCx : 0.470	CCy: 0.406
Average lifetime	10,000 h	

FIGURE 4.57 Emission of light by a metal halide lamp.

lifetime reaches 20,000 h. Figures 4.57, 4.58, 4.60, 4.62, and 4.64 show some commercial metal halide lamps. Figures 4.59, 4.61, 4.63, and 4.65 show their respective emission spectra and Tables 4.20–4.23 list some of the metal halide characteristics.

Their construction is similar to that of other high-pressure lamps, with the internal burner typically made of crystalline aluminum oxide (PCA) due to the use of corrosive elements and compounds, while the outer tube is not coated with phosphors, but acts as a filter of ultraviolet

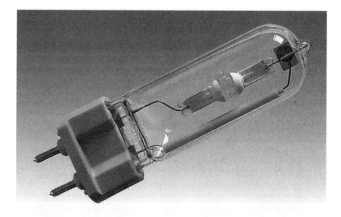

FIGURE 4.58 Metal halide lamp (Philips MHN-TC single-ended Quartz 35 W).

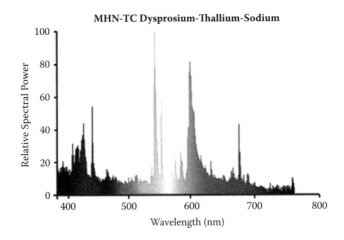

MHN-TC Dysprosium-Thallium-Sodium

FIGURE 4.59   Emission spectrum of a high-pressure metal halide discharge lamp.

radiation emitted by mercury; often, there is a coating to diffuse the light emitted.

4.4.4 High-Pressure Xenon Gas Discharges

The last example in this group of high-pressure lamps is the high-pressure xenon discharge tube. The burner (tube) is made of quartz, and the electrodes are made of tungsten with traces of thorium. The emitted radiation covers the entire visible spectrum, but special attention is required

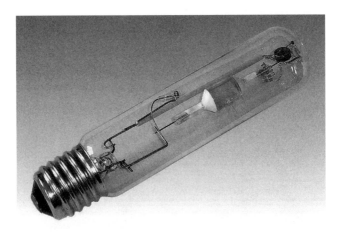

FIGURE 4.60   Metal halide lamp (Thorn Kolorarc MBI-T Tri-band chemistry 250 W).

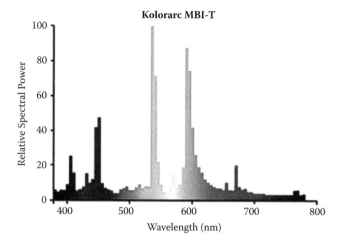

FIGURE 4.61   Emission spectrum of a high-pressure metal halide discharge lamp.

because of the very high pressure and ultraviolet radiation emitted that penetrates the quartz without loss of intensity. The lamps are of low luminous efficacy, but are used wherever white light is required to simulate daylight.

There are several sizes of this lamp in the market, and they are divided into small and large xenon arc tubes ranging in power from a few watts to many kilowatts. By the term *arc*, we refer to high-pressure discharges due to the exact shape given to the plasma by thermal buoyancy. The lamps

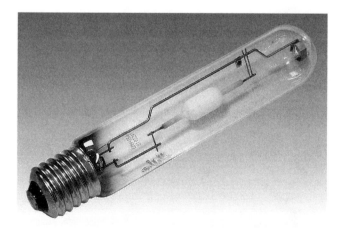

FIGURE 4.62   Metal halide lamp (GE CMH-TT 250 W/830).

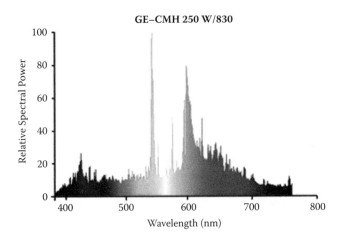

FIGURE 4.63   Emission spectrum of a high-pressure metal halide discharge lamp.

have a high luminous flux, which reaches 200,000 lm under pulsed conditions. Figure 4.66 shows a typical high-pressure xenon discharge lamp.

In a variation of xenon lamps, mercury is added, giving the white light a blue tone and increasing the intensity of UV radiation. Therefore, xenon–mercury lamps are used in applications where ultraviolet radiation is desired for sterilization or ozone formation.

A marriage of technologies, those of metal halide and high-pressure xenon lamps, can be found in automotive lights (Figure 4.67) as mentioned earlier, but they have become known as *xenon lamps* although

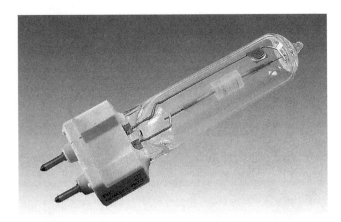

FIGURE 4.64   Metal halide lamp (Philips CDM-T single-ended 35 W).

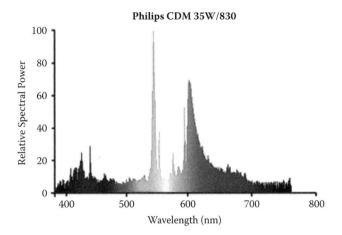

FIGURE 4.65 Emission spectrum of a high-pressure metal halide discharge lamp.

TABLE 4.19 Characteristics of High-Pressure Sodium Vapor Discharges under Different Design Elements

Design Element	Efficacy (lm/W)	Color Rendering Index	Color Temperature (K)	Disadvantages
Simple	60–130	20–25	2000	
Increase xenon pressure	80–150	20–25	2000	Difficult starting
Increase sodium vapor pressure	60–90	60	2200	Lower efficacy and shorter lifetime
Further increase of sodium vapor pressure	50–60	85	2500	Lower efficacy and shorter lifetime
Electronic ballast for pulsed operation	50–60	85	2600–3000	Lower efficacy and shorter lifetime. Special operating gear

TABLE 4.20 Table with Indicative Values and Information about the General Characteristics of a Metal Halide Lamp

Gas Fill	Inner: Ar/Dy, Tl, Na	Outer: Vacuum
Luminous flux	2600 lm (100 h)	
Efficacy	67 lm/W (100 h)	
Color temperature and CRI	CCT: 3800 K	CRI: Ra = 75
Color coordinates	CCx: 0.385	CCy: 0.368
Average lifetime	6000 h	

TABLE 4.21   Table with Indicative Values and Information about the General Characteristics of a Metal Halide Lamp

Gas Fill	Inner: Ar | Hg–In (Na, Tl, Hg)Ix	Outer: N$_2$
Luminous flux	21,000 lm (100 h)	
Efficacy	84 lm/W (100 h)	
Color temperature and CRI	CCT: 4,200 K	CRI: Ra = 70
Average lifetime	6,000 h	

TABLE 4.22   Table with Indicative Values and Information about the General Characteristics of a Metal Halide Lamp

Gas Fill	Inner: Ar/Hg	Outer: Vacuum
Luminous flux	25,000 lm (100 h)	
Efficacy	100 lm/W (100 h)	
Color temperature and CRI	CCT: 3,150 K	CRI: Ra = 85
Color coordinates	CCx: 0.424	CCx: 0.395
Average lifetime	20,000 h	

FIGURE 4.66   High-pressure xenon lamp.

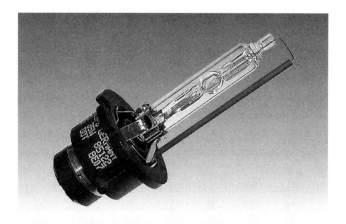

FIGURE 4.67   Metal halide/xenon lamps for car front lights.

xenon serves as the buffer gas and only emits during the first warm-up minutes before the metal halides vaporize. These lamps operate at 40 W and emit about 3000 lm.

4.4.5 Carbon Arc Lamp

The carbon arc lamp combines a discharge with incandescence. The lamp is started by bringing in contact two carbon electrodes so that a discharge is created with a relatively low-voltage application. The electrodes are gradually moved away from each other, and the electric current already flowing maintains the arc. The edges of the electrodes are at high temperature, so they too emit light due to incandescence. Because of their vaporization, the electrodes must periodically be brought closer to each other so that the distance between them remains constant and the arc is maintained. Apart from some specific applications such as video projector screenings, the carbon arc lamps have essentially been retired and replaced by xenon arcs.

4.5 INDUCTION LAMPS

Unlike all other electrical lamps that use electrodes to couple energy (electricity) to the tube, there is a class of lamps where power is transferred to the gas or vapor inductively. The great advantage of induction lamps is that the absence of electrodes gives them much longer lifetimes. In induction lamps one has the possibility to use a greater variety of materials as active media that would not be compatible with electrode materials in conventional tubes.

4.5.1 Low-Pressure Mercury Vapor Induction Lamp

Let us first consider the case of the low-pressure mercury vapor induction lamp. The operating principle is the same with other fluorescent low-pressure mercury lamps as mercury atoms are excited through electron impacts and, after relaxation of the atoms to their ground state, they emit radiation, the ultraviolet (254 nm) resonance line being the main one. The ultraviolet light is then converted to visible light by a phosphor coated on the inside walls of the lamp. The difference here is that there are no electrodes to provide the flow of electrons. Instead, the free electrons in the lamp are set in motion by an electromagnetic field created by an antenna (induction) mounted adjacent to the walls of the lamp. Different companies have presented different setups for this antenna mounting, and some examples can be seen in Figures 4.68–4.70 showing different aspects of the same commercial product, and Figure 4.72 shows another brand. The emission-respective spectra are depicted in Figures 4.71 and 4.72, while the general characteristics of those two induction mercury lamps are listed in Tables 4.24 and 4.25. The electronic ballast generates this high-frequency electromagnetic field, which is transferred through the antenna to the gas. Some common frequencies are 13.6 and 2.65 MHz, that is, in the range of microwaves.

While the average lifetime of electroded fluorescent lamps depends on the average lifetime of their electrodes, the lifetime of the induction

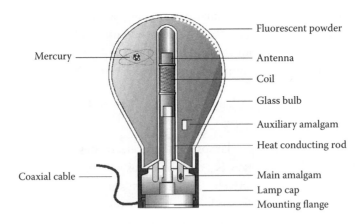

FIGURE 4.68 Anatomy of a low-pressure induction mercury vapor lamp.

FIGURE 4.69 Low-pressure induction mercury vapor lamp (Philips QL Electrodeless Induction system 85 W).

fluorescent lamp is defined by the average life of the operating gear that exceeds 30,000 h.

4.5.2 The Sulfur Lamp

The sulfur lamp is an example of the inductive operating mode allowing the use of an active medium that could not have been used in a discharge tube with electrodes. Here again, a magnetron creates an electromagnetic field at the frequency of 2.65 MHz, which sets in motion the free electrons

FIGURE 4.70 **(See color insert following page 20.)** Photograph of an induction mercury lamp without the phosphor coating.

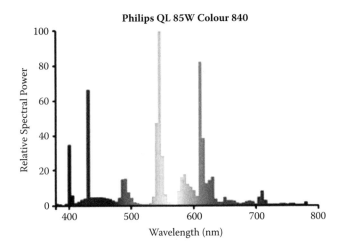

FIGURE 4.71 Emission spectrum of a low-pressure induction mercury vapor lamp.

and raises the temperature of the lamp and consequently the vapor pressure of sulfur (5 atm). Due to the high temperature and pressure, the spherical bulb is made of quartz that can withstand the conditions (Figure 4.74). The luminous flux of a sulfur lamp is quite high (see Table 4.26) compared to other lamp technologies, and lighting a building's interior is possible by directing the emitted light of one lamp to the desired spaces through optical tubes. The spectrum of such a lamp can be seen in Figure 4.75.

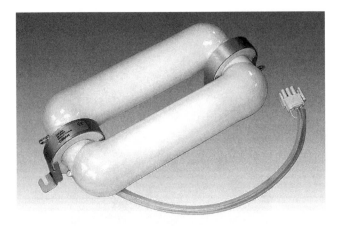

FIGURE 4.72 Low-pressure induction mercury vapor lamp (Osram Endura Inductively Coupled Electrodeless 100 W).

TABLE 4.23   Table with Indicative Values and Information about the General Characteristics of a Metal Halide Lamp

Gas Fill	Inner: Ar-Kr$_{85}$ | Dy-Ho-Tm-Tl-Na-Ix	Outer: Vacuum
Luminous flux	3300 lm (100 h)	
Efficacy	95 lm/W (100 h)	
Color temperature and CRI	CCT: 3000 K	CRI: Ra = 81
Color coordinates	CCx: 0.428	CCy: 0.397
Average lifetime	6000 h	

TABLE 4.24   Table with Indicative Values and Information about the General Characteristics of a Low-Pressure Induction (Electrodeless) Mercury Vapor Lamp

Gas Fill	Ar–Kr–Ne | Hg Amalgam	
Luminous flux	6,000 lm (100 h)	4,200 lm (60,000 h)
Efficacy	71 lm/W (100 h)	49 lm/W (12,000 h)
Color temperature and CRI	CCT: 4,000 K	CRI: Ra = 80
Color coordinates	CCx: 0.390	CCy: 0.390
Average lifetime	100,000 h	

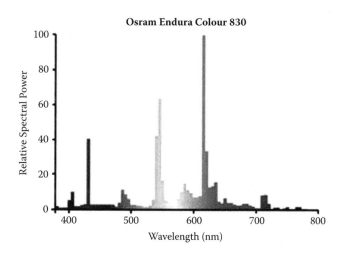

FIGURE 4.73   Emission spectrum of a low-pressure induction mercury vapor lamp.

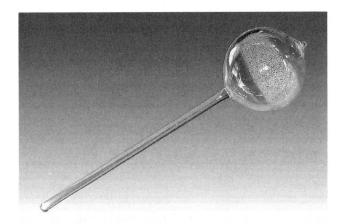

FIGURE 4.74 TUE electrodeless high-pressure sulfur microwave lamp 1000 W.

The emissions are the result of excitations not of sulfur atoms but of sulfur molecules (S2); therefore, the continuous spectrum covers the entire visible region. The luminous efficacy of these lamps exceeds 100 lm/W, and the consuming power is of the order of kilowatts.

The color temperature of this light source is 6000 K, and the CRI is around 80. The spectrum and, therefore, some the characteristics of this white light source, such as the CRI and the color temperature, can be affected or tailored by the addition of other chemicals ($CaBr_2$, LiI, NaI). The average lamp lifetime is defined by the life of the magnetron and reaches 20,000 h. (See Table 4.26 for summary of general characteristics.)

Although not widely used, the sulfur lamp is a light source of very high luminous flux without the emission of ultraviolet or infrared radiation. It is usually employed for lighting large areas either as it is or by using light pipes that distribute the light uniformly over their length and carry the light to many places, thus eliminating the need for other lamps.

TABLE 4.25 Table with Indicative Values and Information about the General Characteristics of a Low-Pressure Induction (Electrodeless) Mercury Vapor Lamp

Gas Fill	Ar–Ne–Kr/Hg	
Luminous flux	8,000 lm (100 h)	5,600 lm (60,000 h)
Efficacy	80 lm/W (100 h)	56 lm/W (60,000 h)
Color temperature and CRI	CCT: 3,000 K	CRI: $Ra = 80$
Average lifetime	60,000 h	

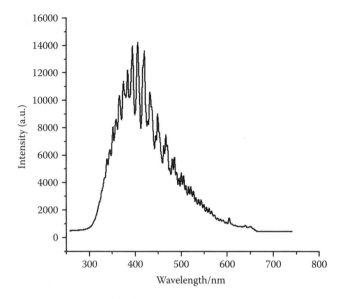

FIGURE 4.75 Emission spectrum of a high-pressure sulfur induction lamp.

Due to the electromagnetic waves emitted with both types of induction lamps, special filters are placed around the bulb to reduce interference with other receivers of this wavelength.

4.6 PULSED OPERATION

The use of pulsed current has opened up new horizons in the field of research and development of novel light source technologies, and some of the new ideas have already taken the road toward the production of new products.

For a discharge lamp of a given power input, the change from a continuous flow of charge (regardless of whether it is unipolar or bipolar) to a pulsed operation (burst of charge flow) means that the average electron energy would shift to higher values, since the applied voltage and current would reach for short periods much higher values for the discharge to be maintained and for the average power to remain the same. This shift in average electron energy would significantly affect the emission spectrum of the lamp. Figure 4.76 shows the different waveforms between the two modes of operation.

Let us examine the case of mercury, which is being employed in the majority of commercial discharge lamps for general lighting.

The emission spectrum of a low-pressure mercury vapor discharge tube under steady or pseudo-steady-state operation (DC or AC current) is

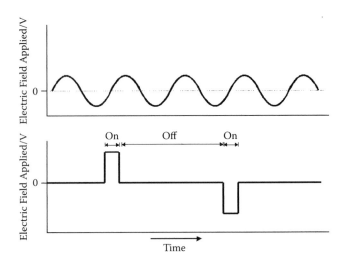

FIGURE 4.76 Waveforms of applied voltage for continuous (AC in this example) and pulsed operation of a discharge tube.

known (Figure 4.77), and the dominant resonance emission line at 254 nm is the one that excites the phosphor on the tube walls and gets converted to most of the visible light output.

Only a small fraction of electrons on the high-energy tail can induce, via collisional excitation, atomic transitions, which would lead to the

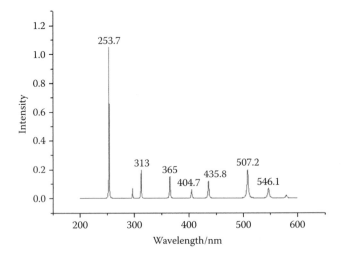

FIGURE 4.77 Emission spectrum of a low-pressure mercury vapor discharge tube under steady-state operation.

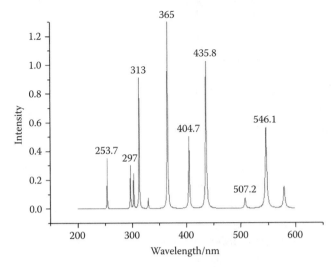

FIGURE 4.78 Emission spectrum of a low-pressure mercury vapor discharge tube under pulsed operation. Instrument sensitivity must also be taken into account when one is considering absolute intensities. The figure here represents relative intensities and how the near-UV and visible lines were enhanced at the expense of the resonant UV line.

emission of near-UV and visible lines and also to ionization, essential for the maintenance of the plasma.

In pulse operation of a discharge tube (pulses of a few microseconds long and frequency of a few kilohertz), the mean energy of electrons is increased; therefore, increasing the ionization degree of the mercury atoms and their excitation to higher-energy states than that of resonance. The spectrum shows an enhancement of the near-UV (313–365 nm) and visible lines at the expense of the resonance line [6], as shown in Figure 4.78. The relative intensities can be controlled by the duration or frequency of the pulses and, consequently, the applied voltage and current waveforms. One possibility arising from this effect is the use of lines close to or in the visible part of the spectrum for phosphor excitation in order to reduce or eliminate Stokes losses.

An important point here is that, during pulsed operation and as the ionization degree increases, the emission of radiation continues in between pulses in the form of afterglow emissions. This afterglow regime is not the same for all emission lines and, as expected, the lines that exhibit the stronger afterglow emissions are the ones mostly enhanced under this mode of radiation. More information and studies on the afterglow effect can be found in scientific publications [6,7].

Details of the electron energies distribution function (EEDF) are very important for the development of models that can predict the behavior of such plasmas, but it is generally assumed that the distribution becomes Maxwellian sometime after the cessation of the pulse.

Another application could be the use of a mixture of phosphors, with each being excited by a different mercury line (e.g., the use of a phosphor excited by the 254 nm line that dominates during steady-state operation in combination with a second phosphor that is excited by the 365 nm line that is prominent under pulsing). The alternation of the operating modes or the control of the pulse duration and frequency would allow the tailoring of the emission spectrum and, therefore, the control and variation of the color temperature or CRI of the source.

The pulsed operation mode was employed in another pioneering work where, for the first time, the focus was *medium-pressure* mercury vapor discharge lamps. The increase of mercury pressure to regimes of a few torr (1–100) and the pulse operation resulted in a white light source without the need to use phosphors. Pulsing and increasing the pressure once again caused an enhancement of the near-UV and visible lines at the expense of the resonance line but, under these conditions, the yellow double line of mercury at 577/579 nm was particularly enhanced (Figure 4.79). This increase in intensity of the yellow line is what causes the source to appear

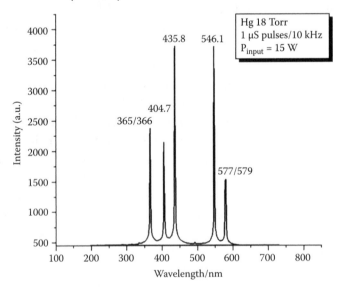

FIGURE 4.79 Emission spectrum of a medium-pressure mercury discharge tube.

FIGURE 4.80 A discharge through pure mercury vapor at pressure of 18 torr without a buffer gas appears white to the naked eye.

white (Figure 4.80), as yellow activates the same eye sensors as red and in combination with the other green and blue lines creates the effect of white light. It is only under the pulsed medium-pressure conditions that the yellow line makes an impact [7].

The adoption of pulse-operating mode will probably not be restricted to discharges of different pressure regimes but will also be used in other technologies. LEDs and incandescent lamps could probably be pulsed in order to control the temperature (of junction or filament) or the response of the human eye.

4.7 ALTERNATIVE DISCHARGE LIGHT SOURCES

It is a fact that the use of pure elements as active media poses the limitation of insufficient vapor pressure in most cases, with few exceptions that have already been tested and are well known. Elements such as the rare gases, mercury, and sodium have already been the base for a number of products, while other elements with useful emission properties can only be used in molecular form and in high-pressure lamps [8], where the temperatures reach high enough values for them to be vaporized and dissociated to a useful degree.

A paper by Zissis and Kitsinelis [9] lists most of the attempts made in recent years to find alternatives to mercury discharge lamps and experimentation from various groups included so far:

- Barium in low-pressure regimes (visible resonance line)

- Nitrogen as a near-UV emitter, oxides of metals (electrodeless configurations to avoid oxygen's reactions with lamp parts)

- Zinc in high-pressure regimes

- Hydroxyl (OH) as a near-UV emitter in low-pressure regimes with rare gas background

- Cyanogen radical (CN) as a near-UV emitter in low-pressure regimes

- Carbon monoxide (CO) in low-pressure regimes as a visible continuous source

- Sodium iodide (NaI), scandium iodide (ScI), indium monobromide (InBr), and other metal halide salts under microwave excitation

In all preceding cases, the efficacies were lower than those of existing discharge lamps and, in some cases, the necessary changes in lamp design and technology constitute further barriers for the development of such solutions.

Especially for the low-pressure cases, all recent attempts to replace mercury focused either on atomic species alone or included the use of molecular compounds but examined the vapor pressures and the emission properties of the molecular candidates and not those of the constituent atoms in case of atomization [10]. The belief that significant atomization occurs and that atoms and diatomics dominate the emission properties of low-pressure discharges regardless of the molecules employed stems from previous experience on mercury-free low-pressure discharges [8,10]. Bearing these extra elements in mind (up to two atoms per species and significant atomization), the author tackled the issue in the past by redesigning the strategy and selection rules [10].

The first criterion is that the new candidates must be efficient emitters in the range of 380–580 nm. The reason for selecting this particular range is the need to eliminate any Stokes losses that would be associated with photons the human eye cannot detect and would need conversion by using a phosphor. On the short-wavelength side, the eye sensitivity, and therefore the visible range, starts at around 360–380 nm. On the long-wavelength side, there was no real need to include wavelengths beyond 580 nm, as sodium is already a strong emitter; just above that limit, the sensitivity of the eye drops and, in any case, the photons on that part of the spectrum cannot be converted by using phosphor. On the other hand, strong emissions in the blue region could be partly converted to cover larger regions of the visible spectrum.

The second requirement is that the emitter be in vapor phase or easily evaporable. Of course, the selected emitter should not be aggressive for the burner envelope and electrode materials. Toxic or radioactive substances also cannot be employed because of environmental considerations.

The search should focus on diatomic species (so as to overcome, at least partly, the vapor pressure limitations of the atomic elements), with at least one of the atoms being an element with useful emissions due to the fact that even with molecules in the gas phase, significant atomization occurs, and the atomic emissions contribute significantly to the overall radiated power output. Finally, one should search for parent molecules if the diatomic species are not freely available or stable in their form [11–13].

Any compound chosen must also be in the lowest possible oxidation state that still vaporizes to the desired degree in order to avoid stoichiometrically driven condensation. What is meant by this statement can be explained with an example. Consider a metal M that vaporizes to the desired degree when in the form of MX4, where X is a halogen or an oxygen atom (or a combination of both as in some oxohalide compounds) but condenses when MX2 forms; then in the plasma phase of a lamp that starts with MX4, all metal will in time condense as it forms MX2 and X2 in the gas phase (such as molecular oxygen or halogen). On the other hand, if the products of the plasma reactions do not form stable species with very low vapor pressures under those conditions, this problem can be avoided.

The author has developed an $AlCl_3$ low-pressure lamp (Figure 4.81 and spectrum in Figure 4.82) [10] after applying a proposed strategy and

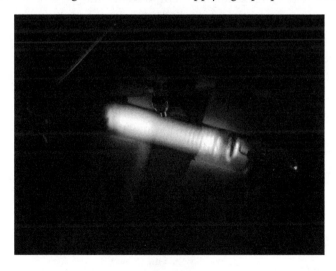

FIGURE 4.81 **(See color insert following page 20.)** Photograph and emission spectrum of a low-pressure metal halide lamp ($AlCl_3$), which has been proposed in the past as an alternative to mercury-filled low-pressure discharge lamps.

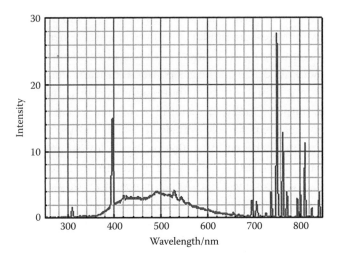

FIGURE 4.82 Photograph and emission spectrum of a low-pressure metal halide lamp ($AlCl_3$), which has been proposed in the past as an alternative to mercury-filled low-pressure discharge lamps.

selection rules. A large molecular band exists between 360 and 650 nm. Overall, the spectrum containing both atomic and molecular emissions justifies the strategy and rules applied.

Having identified promising candidates as active media, the proposed technology of molecular low-pressure mercury-free gas discharges essentially requires the use of the principles of metal halide lamps, in which a variety of spectra can be produced by introducing a mixture of metal compounds and infrared-reflecting coatings used in incandescent and sodium lamps in order to provide thermal insulation and maintain the temperature as high as possible. These coatings could provide the temperature conditions for metal compounds to be vaporized even in low-pressure

TABLE 4.26 Table with Indicative Values and Information about the General Characteristics of the High-Pressure Induction (Electrodeless) Sulfur Vapor Lamp

Gas Fill	S_2—26 mg (5 bar) Ar—75 torr	
Luminous flux	130,000 lm (100 h)	
Efficacy	130 lm/W (100 h)	95 lm/W system
Color temperature and CRI	CCT: 6,000 K	CRI: Ra = 79
Average lifetime	60,000 h (lamp)	20,000 h (magnetron)

and low-power discharge tubes, while an electrodeless tube can solve the problem of the corrosive properties of these metal compounds (see sulfur lamp example).

4.8 MICRODISCHARGE PROPERTIES FOR DEVELOPMENT OF LIGHT SOURCES*

Microdischarges emerged as a new front in the research of nonequilibrium plasmas, introducing a wide range of possible applications, most significantly in the field of light sources, biomedical applications, and nanotechnologies. Development of microdischarges as light sources came forward mainly through efforts to employ alternatives to mercury-based light sources. Additionally, some specified requirements, such as the need to produce thin, large-area homogeneous sources, drew the attention of the nonequilibrium plasma community to the studies of the effects when one downsizes discharges, as well as to the possibilities of relating standard plasmas (centimeter-size and pressures close to 1 torr) to microdimensions and pressures close to atmospheric.

A large number of different configurations of micrometer-size discharges have been studied, but, generally, for the purpose of lighting, large-area flat panels comprising arrays of microcavities proved to be the most promising. As far as the specific geometry of the individual elements of the array is concerned, the hollow-cathode-like structure is accepted as the most favorable and easiest to use in applications. There are several reasons for this. Discharges with hollow cathodes are the most stable, regardless of dimensions. In addition, due to recent advances in processing of materials, fabrication of nano- to micrometer-sized holes in different materials became easier and cost-effective, compared, for example, to parallel-plate discharges that require careful treatment of cathode surfaces and precise alignment of electrodes.

What brought microdischarges to the application field is the possibility of producing nonequilibrium plasmas at atmospheric pressure. Most of the current plasma applications require nonequilibrium conditions. However, nonequilibrium plasmas are produced easily only at low pressures that require high-cost vacuum technology, which is not appropriate for mass production. In standard centimeter-size discharges at atmospheric pressure, the breakdown results in a very rapid growth of ionization, where strong coupling between electrons and ions easily

---

* Section contributed by Dr. Dragana Marić from the Institute of Physics in Belgrade, Serbia.

leads to formation of a thermal plasma. There are several techniques of producing nonequilibrium plasmas at atmospheric pressures or generally at high pressures: by interrupting the discharge temporally or by applying inhomogeneous electric fields; by using dielectric barrier; and by employing small electrode gaps, on the order of micrometers.

Generally, it is difficult to arrive at a good understanding of the operating conditions in microdischarges. Perhaps the best approach would be to start from low-pressure studies and extend them to high pressures and small dimensions by employing scaling laws. Discharges should scale according to several parameters: E/N (electric field to gas density ratio, which is proportional to energy gain between collisions); pd (proportional to the number of collisions); jd^2 (current density times the square of the gap, describing space-charge effects); and ω/N (frequency normalized by gas density for high-frequency discharges).

Most of the applications being currently developed operate in hollow-cathode geometry, and in normal or abnormal glow regime. Recent studies have shown that these discharges do not appear to resemble low-pressure hollow-cathode discharges, and it was found that under the standard experimental conditions the hollow-cathode effect is not significant due to a small electron mean free path. Such results raise the issue of the scaling of discharge properties as a function of pressure (p) and characteristic dimension (d). Since hollow-cathode discharges are complex and have complex characteristics with a number of different modes, it is better to start from parallel-plane microdischarges and follow the scaling over several orders of magnitude of the characteristic dimension d. The disadvantage of this geometry is that it is more difficult to achieve stable operation. On the other hand, the discharge behavior in a simple geometry is easier to understand; it is easier to follow basic processes and their kinetics in gas breakdown and discharge maintenance; and it is certainly the easiest to model.

So far, the studies of parallel-plate microdischarges have shown that, at gaps above 5 μm, standard scaling works as long as the parameters are properly determined and well defined. The two most critical issues in determining adequate scaling parameters, which are commonly disregarded, are the actual path of the discharge and the effective area of the discharge.

Due to a tendency of the discharge to operate at the lowest possible voltage, at high pd-s, it will operate at the shortest possible path between electrodes, while at low pd-s, the discharge will establish along the longest possible path. Under conditions in which the mean free path of electrons is on the order of micrometers, the discharge will tend to operate not only between electrodes but also from the sides and back surfaces, unless that path is blocked. Therefore, effectively, parameter d does not have to be the actual gap between electrodes. This becomes even more important in complex geometries, such as those with hollow cathodes.

Another issue is also associated with the small mean free path in microdimensions. Proper determination of current density (i.e., scaling parameter jd^2) requires proper determination of the effective discharge diameter. The electrode diameter is commonly used as the diameter of the discharge. However, as the discharge diameter depends on the diffusion length, which in turn depends on the electron mean free path, it is obvious that at high pressures the discharge will be quite constricted. Therefore, it is essential to use the actual diameter of the discharge in order to produce the scaling parameter jd^2. Again, in complex geometry, where it is hard enough to determine current density at low pressures, this problem gets even more difficult.

The analysis of volt–ampere characteristics is important to establish the regime of the discharge and also the breakdown voltage. There is an essential difference between Townsend (low-current diffuse), normal glow (constricted), and abnormal glow (high-current diffuse). In glow discharge, there is a cathode fall and possible constriction that indicates importance of space-charge effects. As normal glow has the lowest voltage of all the regimes, it is the most stable and, usually, the point when the transition from normal to abnormal glow is made is the most stable operating point. The large density of space charge leads to a complex field dependence in which mean energy, current, and other properties of electrons are not easily predicted and the range of conditions is limited. On the other hand, in Townsend discharge, where field is uniform and space charge does not cause deviations of the field profile, one can operate at different adjustable E/N, making the discharge more efficient for excitation. Thus, it is favorable to use Townsend discharges for light sources as one could choose conditions where high-energy excited levels are excited efficiently, and thus, one can improve the efficiency of the light source. On the other hand, due to limited and very low current, one cannot produce a large number of photons, so the output is limited. Due to $jd2$ scaling, it is possible in microsize discharges to achieve very large current densities and consequently photon emission for the same $jd2$. However, so far we have not been able to identify the operation of microplasma light sources in the Townsend regime.

Microcavity plasmas have been applied for flat panel light sources with great variation in size, techniques to achieve microdischarges, and possible applications. In addition, similar sources may be used for the production of uniform glow discharges at atmospheric pressure, achieving power densities from tens to hundreds of $kWcm^{-3}$.

Usually, such sources are applied for background lighting applications in both commercial and residential conditions. In the production of these light sources, Al/Al_2O_3 are used. Thicknesses smaller than 1 mm have been achieved.

The microcavities that are used can be made in a number of ways. For example, a diamond may be used with Al foil having characteristic dimension (D) of 250–1600 μm and encapsulated with 10–20 μm of nanoporous aluminum oxide (Al_2O_3). Another interesting way to manufacture microcavities is to use twisted wire bundles, where in the small gaps between the wires a discharge forms with appropriate insulation of wires and applied power.

Arrays with active areas of more than 200 cm² have been reported for operation in the rare gases. Luminance of the order of 1700 cdm^{-2} has been achieved. For Ne/20%Xe gas mixture illuminating a commercial green phosphor, a nonoptimized luminous efficacy of 10.5 lmW^{-1} has been obtained.

Future developments of the microdischarge-based light sources will be in directions of improved or new microcavity geometries and technologies for their manufacture, in optimization of the efficiency and quality of light, and, finally, in developing applications with a wider range of applications.

Open microcavity sources may be used for other technologies, including biomedical for large-area treatment at atmospheric pressure of wounds and for sterilization. Further improvements are required for the luminosity and efficacy of the sources. One way to achieve both would be to use the Townsend regime if possible, but a lot of research is required in this direction before practical success can be achieved. On the other hand, much may be learned from standard-size discharges, and further optimization or at least a better understanding of light source microdischarges may be achieved and should be aimed for.

Thin-panel light sources are yet to achieve the efficacy and luminosity of their larger-size counterparts. Research on scaling and fundamental discharge properties should lead to further improvements in this respect and may be the basis for developing competitive microdischarge-based light sources. In a way, a flat panel plasma TV may be regarded as one such source. Moreover, such discharges may be used as nonequilibrium plasma reactors for a wide range of applications, from physics of materials to biomedicine.

REFERENCES

1. DG—TREN, EU energy and transport in figures, Statistical book 2007/2008.
2. IEA (2006) Light's labor's lost.
3. J.R. Coaton and A.M. Marsden, *Lamps and Lighting*.
4. J. Waymouth, *Electric Discharge Lamps*.
5. M. Jinno, H. Motomura, and M. Aono, Pulsed discharge mercury-free xenon fluorescent lamps with multi-pairs of electrodes, *J. Light Visual Environ.*, 29 (3), 2005.
6. S. Kitsinelis et al., Relative enhancement of near-UV emission from a pulsed low-pressure mercury discharge lamp, using a rare gas mixture, *J. Phys. D: Appl. Phys.*, 37, 1630–1638, 2004.
7. S. Kitsinelis et al., Medium pressure mercury discharges for use as an intense white light source, *J. Phys. D: Appl. Phys.*, 38, 3208–3216, 2005.
8. R. Hilbig et al., Molecular discharges as light sources, *Proceedings of the 10th International Symposium—Science and Technology of Light Sources*, Toulouse, p. 75, 2004.
9. G. Zissis and S. Kitsinelis, State of art on the science and technology of electrical light sources: From the past to the future, *J. Phys. D: Appl. Phys.*, 42, 173001 (16 pp), 2009.
10. S. Kitsinelis et al., A strategy towards the next generation of low pressure discharge lamps: Lighting after mercury, *J. Phys. D: Appl. Phys.*, 42, 2009.
11. R. Payling and P.L. Larkins, *Optical Emission Lines of the Elements*, John Wiley & Sons (2000).
12. K.P. Huber and G. Herzberg, *Molecular Spectra and Molecular Structure IV Constants of Diatomic Molecules*, Van Nostrand Reinhold, New York.
13. R.W.B. Pearse and A.G. Gaydon, *The Identification of Molecular Spectra*, Chapman & Hall, London.

CHAPTER 5

Solid-State Light Sources

O F THE THREE KEY technologies that all artificial light sources in the market are based on, two of them (incandescence and plasma-based sources, otherwise known as discharges) have reached a plateau in efficacies, but nevertheless there is still ongoing research, and fascinating developments continue to take place. The third technology, although it started developing and was marketed much later than the other two, and perhaps for that very reason, is now showing such fast progress that not only can it compete with the other technologies, but in some cases it is preferred to them.

This chapter is an introduction to this technology of solid-state light sources, known as LEDs, which has attracted the interest of many professionals and covers a range of topics answering the most frequently asked questions. Some of the topics include the semiconductor and diode technology on which LEDs are based, ways of creating different colors and white light, modes of operation, their thermal management, applications, and comparisons with the other two technologies of light sources. The chapter addresses a wide range of professionals for whom light and its sources and, in particular, solid-state lamps, are part of their work.

Solid-state lamps, that is, light-emitting diodes, are considered by many to be the future of lighting. Indeed, they have evolved to such an extent that they demonstrate several advantages over the other technologies, and they have already dominated some applications. Of course, there are still some issues that are limiting factors for their further development, such as the issue of thermal management. A large number of scientists are working intensively on the technology of solid-state lamps and, more particularly, they are focused on understanding how to create light through the crystals, the reliability and performance of the materials in order to reduce production costs, the development of phosphor powders for conversion of radiation with good quantum efficiency, the geometry and materials of various parts for better extraction of photons, their sensitivity to temperature and humidity, and, finally, on improving the control electronics of the large number of units needed to produce high luminous flux and color reliability.

There are many who parallel the seemingly inevitable future dominance of LEDs with the case of the transistor, where the hitherto dominant technology of thermionic valve/tube (glass-metal-vacuum) was replaced by a solid-state technology.

Finally, let us not forget that a similar but at the same time different technology of solid-state lighting is based on organic (OLEDs) and polymer (POLEDs) compounds. This technology may prove even more important in coming decades if the materials used become cheaper and more flexible.

5.1 LIGHT-EMITTING DIODE (LED)

The technology of solid-state lighting was the last of the three technologies to penetrate the market, and it is based on the effect of electroluminescence.

The term *electroluminescent* refers to light emission from a solid body when electric current flows through it or when it is placed in an electric field, and it is an effect different from incandescence. The first efforts to create light in this way focused on the use of phosphorescent powder such as ZnS (enriched with copper or manganese) in powder or thin-film form for use as a backlight for liquid crystal displays. These light sources consume little power but require high voltage (>200 V), while their efficiency is low.

These efforts and the development of semiconductor technology that gave birth to the solid-state diode led to a new generation of solid-state light sources.

The controlled addition (doping) of small quantities of certain materials in a semiconductor's crystal structure, such as silicon, without damaging the structure (high-quality crystals must be used, avoiding oxygen and hydrogen) gives the semiconductor some extra properties and, depending on the materials used, we have two different cases. In one case, if the additional material consists of atoms with a number of valence electrons larger than that of the crystal atoms, we call the semiconductor n-type, and it will have a surplus of electrons in the crystal structure (for example, by adding a small amount of phosphorus or arsenic in silicon). In the other

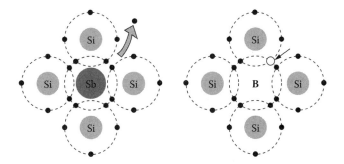

FIGURE 5.1 Creation of semiconductors with additions of various elements to crystalline silicon (doping). The addition of antimony creates free electrons, and the addition of boron creates electron holes.

case, where the added materials consist of atoms with a smaller number of valence electrons (adding a small amount of boron in silicon or gallium), the semiconductor will have a surplus of positive charges, otherwise known as electron holes, and we call such a semiconductor p-type (see Figure 5.1).

The connection of an n-type with a p-type semiconductor creates between them what is called a p-n junction and functions as a diode, allowing the flow of electricity in one direction only, from anode (p-type) to cathode (n-type), as shown in Figures 5.2 and 5.3.

During the flow of electricity through such a solid-state diode, electrons are combined in the semiconductor junction with the positive holes, and this combination puts the electrons in a lower-energy state. The energy

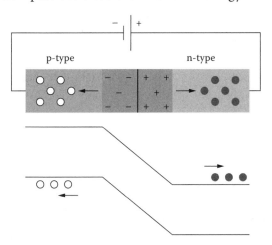

FIGURE 5.2 Creation of a diode by connecting n- and p-type semiconductors.

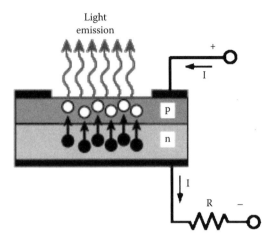

FIGURE 5.3 Radiation emission from the junction of a semiconductor during flow of electricity.

state difference can be released as electromagnetic radiation (not always, as it can also be lost as heat in the crystal) with a wavelength that depends on the materials of the semiconductor. Such a light source is known as LED (light-emitting diode). LEDs emit radiation of narrow bandwidth (a range of a few tens of nanometers). The total charge in the crystal must be distributed as much as possible, and the capsule should be optically transparent (epoxy). The flux is limited by the heat generated at the junction. Both the shell (epoxy) and the crystals begin to wear out after 125°C. Because of the low flux, LEDs have currently limited applications. Due to the materials used, total internal reflection takes place, easily trapping the radiation. The extraction techniques of photons are therefore an important aspect of this technology.

If the material is an organic compound, then we have an OLED (organic light-emitting diode), and in case of a polymer compound the acronym used is POLED.

A brief historical review of the invention and development of this technology follows:

- H.J. Round—Marconi Labs (1907)—pale blue light from a SiC crystal

- Oleg Vladimirovich (1920s)—first LED

- Rubin Braunstein—Radio Corporation of America (1955)—first infrared LED using GaAs, GaSb, InP, SiGe

- Bob Biard and Gary Pittman—Texas Instruments (1961)—first patent for an infrared LED using GaAs

- Nick Holonyak Jr.—General Electric Company (1962)—first red LED

- Jacques Pankove—RCA (1972)—first blue LED using GaN

- M. George Craford (1972)—first yellow LED

- T.P. Pearsall (1976)—first use of an LED in telecommunications

- Shuji Nakamura—Nichia Corporation (1993)—first blue LED using InGaN

- Monsanto Corporation (1968)—first mass production of LEDs, bringing down the cost dramatically

Figures 5.4 and 5.5 show the anatomy of a modern LED unit, while Figure 5.6 shows a variety of shapes and forms that LEDs can take.

The following list is representative of compounds that with appropriate additions of materials (doping) and the connection of the p-type and n-type semiconductors created, result in emission of radiation after flow of electricity. A general rule is that the energy difference increases (wavelength of emission decreases) with increasing aluminum (Al) concentration and decreasing with increasing indium (In) concentration.

AlGaAs—red and infrared:

- AlGaP—green

- AlGaInP—high brightness, orange—red, orange, yellow, and green

- GaAsP—red, orange—red, orange, and yellow

- GaP—red, yellow, and green

- GaN—green and blue

- InGaN—near-ultraviolet, blue—green and blue

- SiC as substrate—blue

- Sapphire (Al_2O_3) as substrate—blue

- ZnSe—blue

- Diamond (C)—ultraviolet

- AlN, AlGaN—ultraviolet

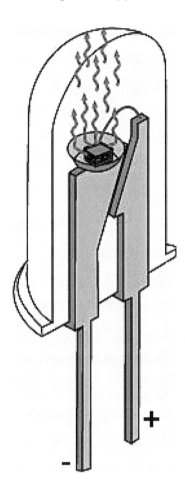

FIGURE 5.4   Light-emitting diode diagram (LED).

Figure 5.7 shows the emission spectra of three different LEDs in three different regions of the visible range. LEDs emit radiation of relatively narrow bandwidth, and this is depicted in Figure 5.8.

The efficiency of LEDs is defined by several factors such as

- The electric efficiency, which has to do with the number of charges in the material (>90% achieved)

- The internal quantum efficiency, which is the number of photons per number of electrons (this depends on the material and construction of layers; heat and reabsorption are the main problems)

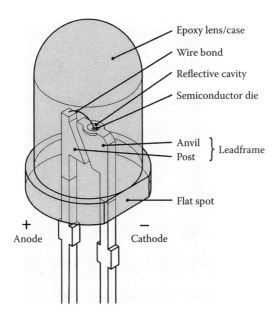

FIGURE 5.5   Solid-state lamp–light-emitting diode (LED).

FIGURE 5.6   Variety of solid-state light sources (LEDs).

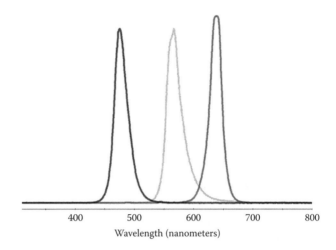

Wavelength (nanometers)

FIGURE 5.7   Emission spectra of three different LEDs in three different regions of the visible range.

- The extraction efficiency, which is the number of emitted photons per total number of photons (the geometry of the material and capsule plays an important role)

- The spectral or optical efficiency, which is related to the eye sensitivity curve (this factor is not taken into account for an LED emitting at the limits of the curve)

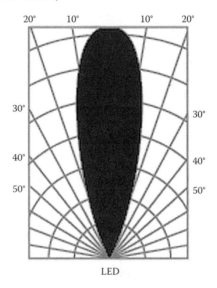

LED

FIGURE 5.8   Light from LEDs is emitted in small solid angles.

5.2 ORGANIC LEDS

If the material used is an organic compound, then it is known as an OLEDs (organic light-emitting diode). For the organic compound to function as a semiconductor, it must have a large number of conjugated double bonds between the carbon atoms. The organic compound may be a molecule with a relatively small number of atoms, in crystalline form, or a polymer (PLEDs), which offers the advantage of flexibility. For the time being, the organic LEDs offer lower luminous efficiencies and average lifetimes than their inorganic cousins.

OLEDs are steadily making their way into commercial devices such as cell phones and flat-screen displays. They are fabricated with layers of organic polymers, which make them flexible, and they use less power and less expensive materials than liquid crystal displays.

The downside is that because the polymers react easily with oxygen and water, OLEDs are expensive to produce—they have to be created in high-vacuum chambers—and they need extra protective packaging layers to ensure that once they are integrated into display devices, they do not degrade when exposed to air or moisture.

OLEDs can be made from a wide range of materials (see Figures 5.9 and 5.10 for examples), so achieving good-quality white light is less challenging. It has not been the quality of light that has let OLEDs down but rather their efficiencies. Fluorescent lighting typically operates at around 60 to 70 lm/W, while incandescent bulbs operate at about 10 to 17 lm/W. In contrast, the best reported power efficiency of an OLED until now has been 44 lm/W.

OLEDs have the potential to grow into a very energy-efficient light source. In production, levels of between 15 and 20 lumens per watt have

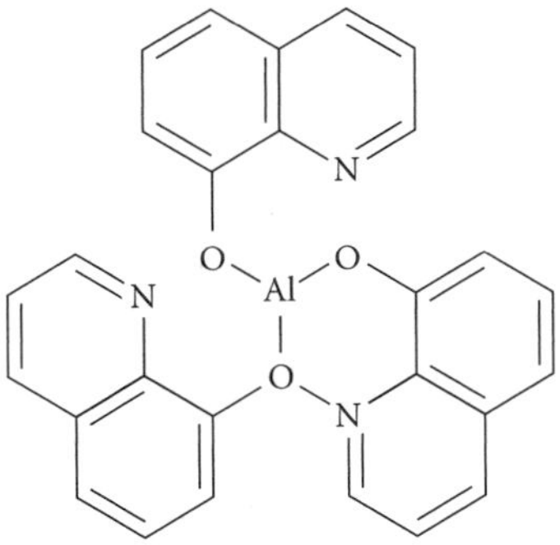

FIGURE 5.9 Chemical structures of organic molecules with double bonds, used for the development of OLEDs.

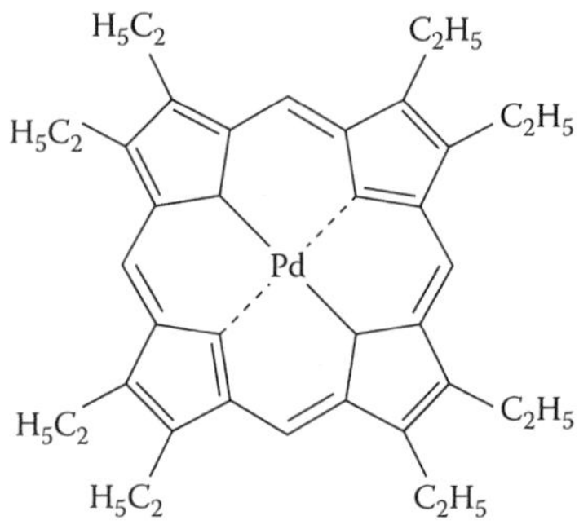

FIGURE 5.10 Chemical structures of organic molecules with double bonds, used for the development of OLEDs.

been achieved, but the ultimate potential is for the technology to reach efficiencies as high as 150 lumens per watt.

A combination of these technologies can also lead to the future light sources. One idea proposed is a hybrid light-emitting diode, or HLED. The device would incorporate both organic and inorganic layers, combining the flexibility of an OLED with the stability of an inorganic light-emitting material.

5.3 LED WHITE LIGHT EMISSIONS

White light can be created with different-colored LEDs (red, green, blue or yellow and blue or four different colors) or by using a phosphor on a UV or blue LED (UV LED with a trichromatic powder or a blue LED with a yellow powder—YAG:Ce).

With three or more primary LEDs (Figures 5.11 and 5.12), all colors can be created. Red LEDs are the most sensitive to temperature and, therefore, corrections need to be made as the LEDs heat up. Moreover, the light intensity and angle of incidence of each LED must match and mix appropriately in order to create the white light correctly. The combination of blue and yellow light also gives the impression of white light since the yellow light stimulates the sensors of the eye that are sensitive to red and green, but the resulting white light will be of low color rendering index.

The other method of creating white light without using more than one LED, is to convert ultraviolet or blue LED light into different colors by using a phosphor. The use of phosphor lowers efficiency due to Stokes losses and other losses on the powder, but it still remains the easiest and cheapest way of creating white light, while the color rendering is usually better due to the larger spectral range of the powder. Another disadvantage

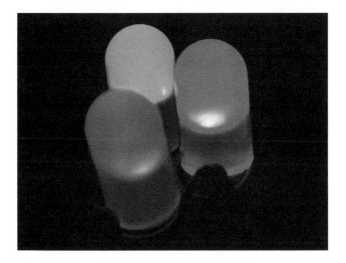

FIGURE 5.11 **(See color insert following page 20.)** Combination of different color LEDs for the creation of white or dynamic lighting.

FIGURE 5.12 **(See color insert following page 20.)** Combination of different color LEDs for the creation of white or dynamic lighting.

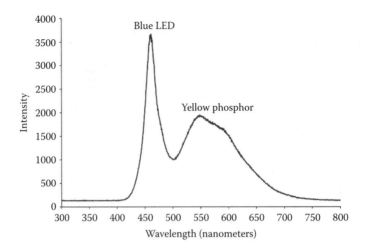

FIGURE 5.13 Blue LED emission spectrum with phosphor that converts part of the blue light into other colors/wavelengths (mainly yellow).

of using a powder is the issue of distribution of light. The light emission angle from the crystal is different from that from the powder; hence, mixing is not very good.

Apart from the UV LEDs that can be used with phosphors, blue LEDs can also be coated with a powder that converts part of the blue light into yellow (Figure 5.13). The ratio of blue to yellow can be controlled through the quantity of the powder used, allowing us to control the color temperature of the source. However, this method gives us a source with a low color rendering index, as there is a deficiency in red emissions, something that can be an issue in some applications such as general lighting but not in other applications such as signage. This method allows us to create white light with a color temperature of up to about 5500 K, but with the addition of another powder that emits in the red part of the spectrum we can also create a warm white light temperature of 3200 K and better color rendering, at the cost of reducing the source efficiency.

The use of powder on a blue LED is the most economical way to create white light, and there are even proposals to use a blue LED (InGaN) with a green powder to replace the low-efficiency green LED or a blue LED with a red powder to replace the temperature-sensitive red (AlInGaP) LED.

The use of phosphors on an UV LED can give white light of significantly higher color rendering index, but at the expense of efficiency (mainly due to Stokes losses), similar to how a fluorescent mercury lamp operates. The different powders must be coated in such a way that there is no absorption of each other's emitted light. The powders used in fluorescent lamps are not appropriate as they are stimulated by the mercury emission lines at 185 and 254 nm, while the UV LEDs emit at 360–460 nm.

With three LEDs, we have better control over color (dynamic lighting), while the use of phosphors gives stability and a better mix. There are, of course, products that use two or three crystals in the same LED with appropriate wiring in order to create different colors and have better color mixing and control without the need for phosphors. This technology, however, of many crystals in the same LED raises the cost due to the separate control of each diode that requires more gear.

Whether one uses UV or blue LEDs with appropriate phosphors or suitable semiconductors, a variety of colors and accents of white can be produced today according to market demands.

To use three LEDs (each primary color) to create white light means that they have to be controlled during operation as they wear out differently and show different sensitivities to heat. The appropriate electronic and optical components can provide this control. When using phosphors, one cannot control or make corrections, and the increase in temperature shifts the emission wavelengths of blue LEDs.

The following table (Table 5.1) lists the advantages and disadvantages offered by the different ways of creating white light with LEDs.

A third method of producing white light without the use of fluorescent powders is through combination of radiation simultaneously produced by the semiconductor and its substrate (blue radiation from ZnSe and yellow

TABLE 5.1   Comparisons of Different Methods for Creating White Light

	Advantages	Disadvantages
Mixing different color LEDs	Dynamic lighting Ability to create millions of colors Better efficacy Control of component colors	Different colors have different sensitivities to heat/no stability Complex electronic gear Not good color mixing
Blue LED with phosphor	Good efficacy Good color rendering index Wide color temperature range Better stability	Not good color mixing at certain angles No control or regulation of different colors
Ultraviolet LED with phosphor	Good color mix Wide color temperature range Good color rendering index Better stability	Poor efficacy Low power Must manage UV light No control or regulation of different colors

light from the ZnSe substrate). The absence of fluorescent powder means higher efficacy.

Finally, a method that is being developed rapidly is the use of quantum dots: nanocrystalline semiconductor materials with dimensions equal to a few dozen atoms that emit light (fluoresce) with high efficiency under electrical or optical stimulation. The wavelength of radiation can be controlled by controlling the size of the nanocrystals, and this method is in the experimental stage (see Figure 5.14 for related spectra).

This color-tailoring ability solves one of the major problems with using LEDs for general lighting applications. LEDs are appealing because they last for years, use perhaps 20% of the electricity of a standard incandescent bulb, and are highly efficient at converting electricity into visible light instead of into heat. But to make white light, you either have to mix together LEDs of different colors or use a blue LED coated with a phosphor that emits yellow light

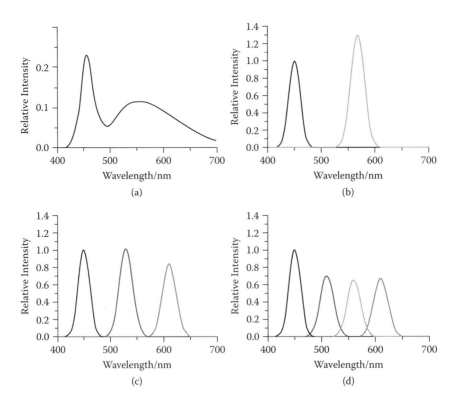

FIGURE 5.14 Various emission spectra of white LEDs: (a) blue LED with phosphor while the other three spectra, (b), (c), and (d), belong to dichromatic, trichromatic, and tetrachromatic quantum dot LEDs, respectively.

FIGURE 5.15   An LED lamp comprising many LED units.

to produce a whitish mix. The problem with the phosphors is that they do not emit evenly across the visible spectrum. They tend to have gaps in the green section and even more so in the red, leading to the harsher, bluish light.

Due to their low levels of emitted light, a lamp must make use of multiple LEDs in order to be functional as a light source for general lighting. Many LEDs are needed for most applications (Figure 5.15).

Most LEDs on the market operate at low power, usually less than 1 W, but some products operate at powers as high as 7 W with an efficacy of 20 lm/W. The highest efficacy ever reported is about 130 lm/W, but for very low-power LEDs. Generally, LEDs emit from 1 lm to several tens of lumens, while there is now on the market a 5 W LED 5W with 120 lm of white light at 350 mA. Depending on their flux, LEDs are categorized as follows:

- Indicators with <5 lm

- Standard with 5–50 lm

- High Brightness (HB) with 50–250 lm

- Ultra High Brightness (UHB) with >250 lm

If LEDs reach 250 lm/W in the next two decades, they could replace all fluorescent lamps that are currently limited to 50–120 lm/W.

A general rule that has been stated and that so far seems to be holding is the doubling of light output from each LED every 24 months in the last 40 years (similar to Moore's law) and the halving of cost almost every decade. This pattern was observed and described for the first time by Roland Haitz. In most cases, LEDs operate with less than 100 mA, and when the current exceeds that value, they are referred to as power LEDs. For high-power LEDs, a silicon gel is used instead of a polymer epoxy coating, and an appropriate heat sink is used for proper heat management.

5.4 LED OPERATION

The LED is a diode; that is, electric charge flows only in one direction. Once the LED starts to operate normally, the current is linearly proportional to the voltage; that is, the relationship between voltage and current for the LEDs is positive, as shown by Figure 5.16.

$$V_f = V_o + R_s I_f$$

V_o is the initial voltage that must be applied before charges start to flow. The value depends on the material and the energy difference between states (band gap). In general, the voltage is influenced by the temperature of the diode. The resistance also depends on the material and is quite low.

During production, there are shifts in V_o and the resistance, so the categorization of the final products according to their electrical characteristics (binning) is necessary as the light output is proportional to the intensity of the current and, therefore, the voltage.

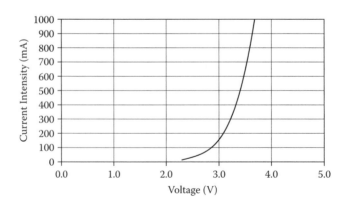

FIGURE 5.16 Voltage–current relationship in LED starting and operation.

The two most important factors in limiting light emission are heat and current density. At large currents, the materials wear out, mainly due to high temperatures but also due to leakage of charges to the other layers outside the active crystals (especially when the layer of the active crystal is thin). The maximum values depend on the material but, in general, more than 120 A/cm^2 is prohibitive for all cases. High temperature not only damages the crystal but also causes shifts in the emitted spectrum. For example, yellow can be converted to red at high temperatures just before the destruction of the crystal and the diode junction. This change corresponds to about 0.1 nm per degree Celsius, something that is a great disadvantage if color reliability is important.

LEDs generate heat during operation (not in emission), and the materials are very sensitive to it. Thus, the thermal design is an important part of the design of these light sources.

An LED is a low-power light source, so LED lamps cannot be connected directly to the mains voltage as they need a current and voltage controller to keep them at low values and also to be able to run several LEDs at once (as shown in Figures 5.17 and 5.18). In other words, LED lamps need the equivalent of ballast, which we call a driver.

The driver can be a constant voltage driver or a constant current driver.

The constant voltage driver is not as reliable because there is a differentiation of the required voltage during production; and because the resistance is small, variations in voltage result in significant variations in intensity.

The constant current driver is preferable as the intensity is set to the normal operating intensity of the LEDs, and the voltage is adjusted.

There is also a hybrid driver of constant voltage with a large resistor parallel to the LED. However, the power consumption by the resistor renders such drivers inefficient.

A driver can be used for many LEDs, and it is important how they are connected (series/parallel). If we have a constant voltage driver, then a resistor should be added so that we have nominal current intensity, while with constant current drivers, we need to take into account the number of LEDs. In each case, the current intensity through each LED or series of LEDs must be carefully estimated.

The position of each LED in a system plays an important role in the functioning of the whole layout. If an LED fails, it can create an open circuit (e.g., in a series connection) or a closed circuit with different electrical

FIGURE 5.17 **(See color insert following page 20.)** LEDs in series mainly for decorative lighting.

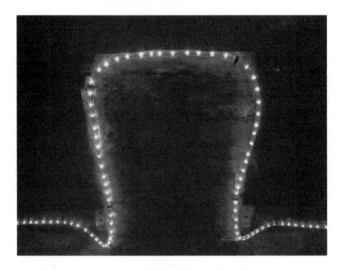

FIGURE 5.18 **(See color insert following page 20.)** LEDs in series mainly for decorative lighting.

characteristics (e.g., if all parallel) such as higher current in one of the series leading to asymmetric production of light.

A combination of series and parallel connections offers greater reliability because the current can find alternative paths without overloading some units.

Protection from large positive (normal flow) and negative (opposite polarity) voltages is provided by high-voltage diodes in appropriate connections. LEDs withstand very short (<1 ms) and nonrecurring high-current pulses (hundreds of milliamperes).

Of course, there are intelligent LED control systems that cannot only create a wide range of colors by combining multiple light sources and offer dynamic lighting, but can also take into account variations in temperature or other electrical characteristics and make the appropriate corrections and changes. There are many communication protocols to control lighting systems and the choice depends on the application. It is important for an intelligent LED driver to be able to receive and analyze the following list of signals:

- Linear voltage control (0 to 10 V)

- Digital multiplex—DMX512

- Digital addressable lighting interface (DALI)

- Power-line communication (PLC)

- Domotic standards: INSTEON, X10, Universal Power-Line Bus (UPB), and ZigBee

5.5 THERMAL MANAGEMENT OF LEDS

All light sources convert electrical energy into heat and radiation emitted in varying ratios. Incandescent lamps emit mostly infrared radiation and a low percentage of visible light. Low- and high-pressure discharge lamps (fluorescent and metal halide, respectively) produce more visible light but also emit infrared and ultraviolet radiation as well as heat. LEDs do not emit infrared and undesirable ultraviolet radiation and apart from visible light produced, the remaining energy is converted into heat, which must be transferred from the crystal to the circuit (as the capsule surrounding the LED is not thermally conductive, the heat flows in the other direction), and from there to

TABLE 5.2 Power/Energy Conversion for Different "White" Light Sources

	Incandescent (%)	Fluorescent (%)	Metal Halide (%)	LEDs (%)
Visible light	10	20	30	15–25
IR	70	40	15	~0
UV	0	0	20	0
Emitted energy	80	60	65	15–25
Heat	20	40	35	75-85
Total	100	100	100	100

other parts of the unit, until finally it is transferred to the environment through the air. See Table 5.2 for a comparison of different technologies regarding energy conversion.

The removal of heat happens first with its flow from part to part and then from the surfaces to the environment. This means that many factors play a role in how efficiently and quickly the heat is removed, and some of these factors are as follows:

- The materials of which different parts are made (preferred materials are those that are thermally conductive, that is, having low thermal resistance such as metals, but new polymers also exhibit good conductivity)

- Their connections (there should be a good contact between each part, and gaps do not help because air is not a good conductor of heat)

- Their total surface area (large areas with as low a volume as possible)

Heat production (and the temperature increase at the diode junction which accompanies it) is the major limiting factor and the biggest obstacle to developing LEDs of higher power and brightness. Therefore, the issue of thermal management is currently perhaps the most important problem that scientists and technologists have focused on, and their efforts are directed toward finding and using material of high thermal conductivity to reduce as much as possible the thermal resistance R of the system so that heat is removed as quickly and easily as possible.

The thermal resistance R_θ of an LED is defined as the ratio of the temperature difference between the junction and the environment (ambient temperature) over the consumed power due to the current that flows through the LED.

$$R_\theta = (\Delta T_{junction-ambient}/P)$$

Where

$\Delta T = T_{junction} - T_{ambient}$
$P =$ current intensity (I) * voltage (V)

The total thermal resistance must, of course, include the entire system from the junction to the surfaces that are in contact with the surrounding air. Each individual part of an LED is characterized by a different thermal resistance (whose values depend on the geometry, material, and surface area of each piece), and according to these we can define the thermal resistance of the crystal, which is given by the manufacturers (the smaller the value, the easier the transfer of heat). Figure 5.19 shows a thermal model for an LED circuit that is analogous to an electrical circuit.

With the thermal resistance (obtained from the manufacturer) between the junction and the material on which the crystal rests, and by measuring the temperature differences between the other parts using infrared

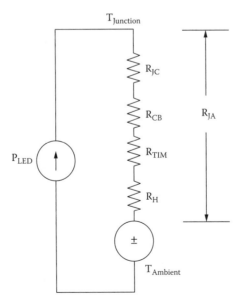

FIGURE 5.19   Thermal model of an LED "circuit."

detectors or thermocouples, we can calculate the temperature at the junction for different current intensities and power values.

The thermal resistance, with units °C/W, is deduced from the thermal conductivity (with unit W/mm), the length of the heat conductor, and its cross-section. This practical parameter allows us to calculate various temperatures at different parts of the system when the consumed power is known. The model used is that of an electrical circuit, where the parallelisms are

Heat Q (W) ~ Current intensity

Thermal resistance Rθ (°C / W) ~ Electrical resistance

Temperature difference ΔT (°C) ~ Voltage

The equivalent to Ohm's law is $\Delta T = Q \times R\theta$

5.5.1 What Defines the Junction Temperature?

Light and heat are produced at the junction of the diode, which has small dimensions, so the heat production per unit surface area is very large. The temperature of the junction cannot be measured directly, but it can be calculated by measuring the temperature of another part and taking into account the thermal resistances of all materials.

There are three factors that determine the junction temperature of an LED: the intensity of the operating current, the thermal resistance of the system, and the ambient temperature. Generally, the greater the intensity of the current, the more the heat produced in the crystal. The heat must be removed in order to maintain the flux, the lifetime, and the color. The amount of heat that can be removed depends on the temperature and the thermal resistances of the materials that make up the whole LED.

The products on the market have a maximum temperature at which they must operate, which is around 120°C. The efficacy and lifetime, however, begin to decline well before that temperature limit. Very few power LEDs have the appropriate initial design that allows them to function at maximum power without using a secondary cooling system. Temperature increase without proper control and stabilization is certainly the main reason for early destruction of LEDs. Although 120°C is given as the maximum operating limit, a more realistic limit is 80°C as one must take into account that fluctuations of the ambient temperature can be of the order of 25°C or higher.

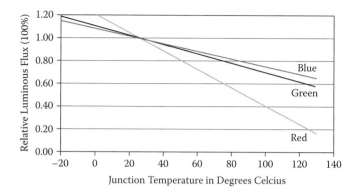

FIGURE 5.20 Effect of temperatures on lifetime and flux for different-colored LEDs presented for known commercial products.

Manufacturers classify LEDs according to the luminous flux and color under pulsed current (25 ms pulses), keeping the junction temperature constant at 25°C. But under normal operating conditions, the junction temperature is at least 60°C, so the flux and the color will be different from those of the manufacturer's specifications. The worst aspect is the LED thermal management; the greater are the differences and the efficacy losses.

The rise in junction temperature has various consequences such as reductions in efficiency, life expectancy, and voltage value as well as shifts of the radiation wavelengths; the latter, especially, affects the operation of mostly white LEDs, causing changes in color temperature (see Figure 5.20).

In general, temperature affects each color to a different extent, with red and yellow LEDs being the most sensitive to heat and blue being the least sensitive. These different responses of each color to temperature can lead to changes and instabilities of the white light produced by RGB systems if during operation the junction temperature T_j is different from that specified by the manufacturer. But even when phosphors is used, the shifts are significant because the powders are sensitive to specific wavelengths.

5.5.2 Overheating Avoidance

There are two ways in which we can control the junction temperature to avoid a malfunction or premature destruction of an LED. One is to decrease the intensity of the current, so that the LED is operated at a lower power, and the other is to carefully design so that the overall thermal resistance of the LED is minimized. The second method, although preferable, is not so simple and requires us to take into account several parameters such as the total surface area of the materials in contact with the surrounding air. The greater the contact area, the lower the thermal resistance.

A combination of both methods is common practice, but proper calculations must be done in each different case. Since the flux decreases with both increasing temperature and decreasing current, and thus the power, we must work out whether (taking into account the thermal resistance given by the manufacturer) a decrease in power, which means operation with lower brightness, ultimately pays. The voltage also depends on the junction temperature, and it is reduced as the temperature increases. If we deal with a single LED or several in a series, then one can control the current; otherwise, there is a risk of having uneven distribution of the current intensity due to temperature variations and, ultimately, destruction of the system.

The circuit connected to the LED may include some type of heat sink, such as the MCPCB LEDs (metal-core printed circuit board), but higher-power commercial LEDs are using extra heat sinks characterized by their large size relative to the LED and their large surface area. A typical geometry of such extra large heat sinks includes the use of fins that dissipate heat more efficiently. Such a heat sink is incorporated in luminaires or can be itself the luminaire. Finally, some LEDs have the circuit at a distance so that the heat generated by the circuit does not contribute to the increase of the junction temperature. See Figures 5.21 and 5.22, which depict LED lamps with incorporated heat sinks.

FIGURE 5.21   LED lamps with heat sinks incorporated in the design.

FIGURE 5.22   LED lamps with heat sinks incorporated in the design.

In addition to electrical, mechanical, and optical data, a compact model for Flotherm software is now permanently available at www.osramos. com/thermal-files for calculating thermal behavior. Customers can find all the documents and the latest data needed for calculating thermal variables for different designs without building costly prototypes or carrying out time-consuming measurements. The data are available for standard high-power LEDs in the visible range, particularly for the Dragon family, the Power TopLED range, the Advanced Power TopLED range, and some Ostar variations.

The compact model available on the Web site is a simplified thermal geometry model that can be integrated in Flotherm software and used for customer-specific calculations. It is suitable, for example, in calculating the temperature distribution in a planned system.

5.6 DIMMING/CONTROLLING THE BRIGHTNESS

An LED can be dimmed by controlling the current, and this can be done in two ways:

- Increasing/decreasing the intensity (DC dimming)
- Pulse control

Controlling the DC current intensity has some disadvantages, which are described in the following text. Binning of LEDs takes place under the operation current, so by changing the intensity there is no longer reliability regarding common features between LEDs of the same category (proposed dimming to ¼ of the intensity only). Low intensity also implies big changes in parallel LEDs due to changes in voltage and, finally, changes in intensity can lead to differences in color.

On the other hand, with pulse control, the maximum pulse current intensity value is set at the normal operating value and, thus, no changes occur in the characteristics, while the mean or average intensity is defined by the frequency and duration of pulses. In this way, the LED works properly, and a linear relationship between current intensity and luminous flux is ensured. For controlling the duration of the pulses, this is an easy task because LEDs respond instantly ($<\mu s$).

The dimming ratio is defined as the minimum mean intensity value over the maximum mean intensity of LED current. This percentage is determined by the shortest possible pulse that the driver can deliver and which at its maximum reaches the nominal current intensity of LED operation. The shorter pulse is in turn defined by the rise and fall times of the pulses.

This method makes dynamic lighting possible, and it is imperative that the minimum frequency chosen should be one at which the source is comfortable to the human eye.

Of course, with a combination of pulse control and DC current intensity checking, even smaller dimming ratios are possible.

5.7 GENERAL CHARACTERISTICS OF LEDS

A summary of the general characteristics of LEDs follows (see also Table 5.4):

- Their emission spectra are narrow band (a few tens of nanometers).

- They are characterized by their low brightness (flux), starting from 1 lm for conventional LEDs and reaching 120 lm for high-power LEDs.

- Their efficacy is around 20–30 lm/W, which already exceeds that of incandescent lamps but lags behind efficacies of discharge lamps such as low-pressure fluorescent and high-pressure metal halides.

- For high flux, many LED units are necessary.

- Wide range of colors with RGB mixing.

TABLE 5.3 As the Temperature Rises, Wavelength Shifts are Different for Different-Colored LEDs

Color	K (nm/°C)
Yellow	0.09
Red	0.03
Blue	0.04
Green	0.04

- They respond instantly without the switching frequency wearing them out. LEDs with phosphor have a slightly slower response time due to the powder fluorescence.

- Good brightness control and dimming rations reaching 1/3 with current intensity decrease and 1/300 with pulse control (a few hundred hertz).

- Their end of life is characterized by a gradual decrease of the luminous flux and is not sudden.

- They are sensitive (and wear out if exposed) to heat and static electricity, and, in the case of blue and ultraviolet LEDs, to radiation. Generally though, they are characterized by their material strength in contrast to the fragile technology of glass–gas.

- They are characterized by their small size, which means freedom in luminaire design. For large fluxes, however, the luminaire must also be large.

- Their average lifetime (70% of initial lm value) is 50,000 h. The end of life depends on defects in the crystalline structure of the semiconductor or the fluorescent powder.

TABLE 5.4 General Characteristics of LEDs

Efficacy (lm/W)	<130, most around 20–30
Power (W)	0.1–7
Color temperature (K)	Wide
Color rendering index	Up to 90
Lifetime	50,000–100,000
Applications	Signage, remote control, fiber-optic communication, decoration, advertising,

- There is no harmful radiation in the ultraviolet and infrared parts of the spectrum, which is an advantage compared to other lamps, but sensitivity and reduced efficacy at elevated temperatures caused by the current flow are disadvantages. Almost all LEDs have an upper limit of 125°C.

- LEDs are still expensive, not just because of their materials and manufacturing cost but also due to their low luminous flux. Nevertheless, they offer a low power consumption solution in applications that require low levels of light.

5.8 APPLICATIONS

Applications in which LEDs have the advantages are

- Applications that require light of specific color. LEDs produce specific colors more efficiently than by putting filters in incandescent lamps.

- Applications in which long lamp lifetimes are required due to difficulty or high cost in replacements.

- Wherever small-sized light sources are required, such as decorative lighting or small spaces (mobile phones, car interiors, etc.).

- Wherever instant start and dimming are necessary.

Applications where LEDs have disadvantages and must not be widely used yet are the following:

- In high-temperature environments

- Where high brightness is required

- Where color stability is necessary

- If accurate stability of color temperature and color rendering index is essential

- If good knowledge of lifetime and lumen depreciation is needed

The goal of LED companies is for this light source technology to dominate the following applications in the near future:

- TV and computer large screens

- Small projection systems

- Car headlights

- Interior lighting

Organic/polymer LEDs (OLEDs/POLEDs) will also play an important role here, but for now they are characterized by short lifetimes.

The different applications can be categorized into those where visual contact with the source is necessary (signage) and those where the reflected light is used (general lighting).

5.8.1 Signage—Visual Contact

In applications in which high levels of brightness are not needed but the creation of optical signals of specific color is the aim, LEDs have already started dominating the market. Some of those applications are the following (LED characteristics that offer the advantages are in brackets):

- Traffic lights (color, sturdiness)

- Car back lights (style, size, electrically compatible) as shown in Figure 5.23

- Car interior (size, mercury-free)

- Decorative lighting (size, dynamic lighting, dimming) as shown in Figure 5.24

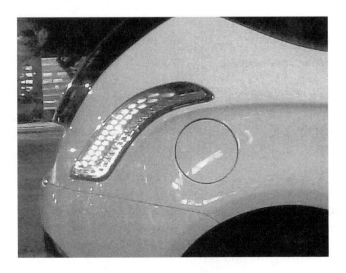

FIGURE 5.23 Use of LEDs for automotive brake lights.

FIGURE 5.24 Decorative lighting with strips of red LEDs.

- Monitors and screens (color mixing)
- Road signs (long lifetime, size)
- Mobile phones (size, low voltage)

5.8.2 General Lighting—No Visual Contact with the Source

There are some applications of general lighting where LEDs are preferred for various reasons:

- Dental treatments/bleaching (blue color, size, long lifetime, replacement of low-efficiency halogen lamps)
- Torches (size, low voltage)
- Architectural lighting

- Machine sensors (various geometries, durability, sturdiness)

- Telecommunications (instant start)

LEDs are used in a wide range of applications ranging from building lighting to large panel displays.

For LEDs to dominate in other markets, the following characteristics must be improved or changed:

- Crystal materials and layer geometries so that efficacy is increased (internal and extraction)

- The materials of all parts of the unit so that they withstand higher temperature; thus, higher power inputs in order to increase the brightness/luminous flux

- Cost, which can be lowered with the help of the aforementioned changes

- Better quality of white light with higher color rendering index and wider range of color temperature

- The production line, so that there is greater reliability and no variation in efficiency

5.9 CLOSING REMARKS

The future of LEDs and perhaps of lighting in general will depend on the outcome of various research paths that scientists around the world are following or must follow. It is essential that a better understanding of the light-generation mechanism and improvements of internal quantum efficiency be achieved. LED manufacturing techniques must also improve so that consistency and better quality can lead to better marketing. Apart from the actual junction, other parts of an LED are also crucial and require R&D, such as substrates, packaging, and lenses with proper thermal management. The ultimate goal is not only to increase the efficacy of each LED but also to achieve long lifetimes and tolerance to high temperatures. In addition, it is important to develop phosphors capable of absorbing and converting photons efficiently and, as is the case already for fluorescent lamps, the development of "quantum-splitting phosphors" could be a breakthrough for LEDs.

The technology of OLEDs also has the potential for rapid growth and market penetration as the great variety of organic luminescent materials could give a large number of emitting colors.

As with any other technology, high efficiencies, consistency, long lifetimes, good color stability, uniformity over large surfaces, and relatively low costs are the characteristics desired for a technology to succeed.

REFERENCES AND USEFUL LINKS

- Schubert, E.F., *Light Emitting Diodes*, Second edition, Cambridge University Press, 2006. United Kingdom.

- A. Zukauskas, M.S. Shur, and R. Gaska, *Introduction to Solid State Lighting* (Wiley, New York, 2002).

- www.LightEmittingDiodes.org.

- www.lumileds.com.

- www.lrc.rpi.edu/researchareas/leds.asp.

- www1.eere.energy.gov/buildings/ssl/basics.html.

- scitation.aip.org/journals/doc/PHTOAD-ft/vol_54/iss_12/42_1.shtml.

- www.netl.doe.gov/redirect.

- www.maxim-ic.com/appnotes.cfm/appnote_number/1883.

- www.ecse.rpi.edu/~schubert/Light-Emitting-Diodes-dot-org.

- www.intl-lighttech.com/applications/led-lamps.

- ieee.li/pdf/viewgraphs_lighting.pdf.

- http://trappist.elis.ugent.be/ELISgroups/lcd/tutorials/tut_oled.php.

- http://www.ewh.ieee.org/soc/cpmt/presentations/cpmt0401a.pdf.

STANDARDS

IESNA LM-79-08

ANSI C82.2 (efficacy)

IESNA LM80-08 (lumen depreciation)

ANSI C78.377A (CRI)

Test standards

ANSI C82.77-2002 (PFC)

EN61000-3-2 (harmonics)

EN61000-3-3 (flicker and voltage variations)

CHAPTER 6

Dynamic Lighting

W HILE ONE CAN CONTROL the levels of brightness of a light source through dimming, other characteristics such as the color temperature or the color rendering index are fixed, so for each different application a different lamp is chosen. Dynamic lighting is a new field of lighting system development that enables one to control parameters such as the color temperature or the color rendering index of the system. With dynamic lighting, one can create a lighting environment that can be changed at will to match the mood or activity of people in that environment.

Dynamic lighting is

- Creation of an atmosphere appropriate to the activity

- Creation of an atmosphere appropriate for a certain mood

- Simulation of natural light that is not static but changes during the day and generally over time

- Creation of different esthetic effects

The main rule in dynamic lighting is that a cool white light creates a stimulating and active environment (6000°C), while a warm white light creates a relaxing and pleasurable environment (3000°C).

Here are some examples of different environments, functions, and atmospheres that dynamic lighting can create:

- **Meetings**—Medium level of brightness and high color temperature for stimulation

- **Focused work**—High color temperature and high brightness levels for office work that demands focus

- **Relaxation**—Low brightness levels and low color temperature for resting or hosting guests

- **Product promotion**—Different tones of white light enhance the image of certain products, so, for example, winter products would appear more attractive under a cool white, etc.

Here are some examples of desirable changes during the Working Day:

- **Morning**—Higher color temperatures and brightness levels stimulate and boost activity

- **Midday**—Lowering of brightness levels and color temperatures to create a relaxing environment

- **Afternoon**—This is the time of day when most people feel sleepy or tired. An increase in brightness levels and color temperatures would lift activity.

- **Evening**—For all those working late, a warm white light can create a pleasant atmosphere.

- **Daylight simulation**—The mood of people working in closed spaces with no natural light (no windows) can be affected by the monotonous static lighting. Dynamic lighting systems can offer a solution by creating a changing atmosphere simulating the changes of natural light during the day.

There are some products on the market that give the user the option of controlling and changing the color temperature. These products make use of multiple light sources, with some of them emitting warm white light (i.e., 2700 K) and others a cool white light (i.e., 6500 K). By dimming different lamps, the user can control the overall color temperature of the

FIGURE 6.1 **(See color insert following page 20.)** For the creation of all colors, one must use three light sources in a system.

system. So, for example, a cool white light can be chosen for a task that requires focus and a warm white light would create a more comfortable environment suitable for relaxation. By proper dimming, the whole color temperature range can be achieved.

Moreover, if one uses three lamps each emitting light of one of the three primary colors (red, green, blue), as shown in Figure 6.1, then any other color can be created through mixing of the primaries, and space lighting can be done in many ways (Figures 6.2 and 6.3).

The disadvantage of using multiple light sources for the creation of varying lighting environments is that expensive optics are needed in order to mix and diffuse the light from each source and create a homogenous final emission. During the process of diffusion and mixing of light

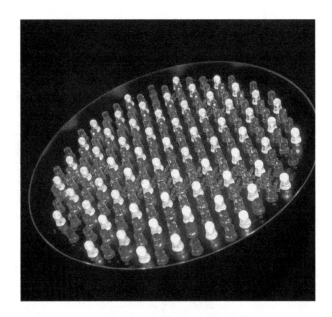

FIGURE 6.2   **(See color insert following page 20.)** Dynamic lighting can be achieved with the use of multiple colored sources in the same system.

FIGURE 6.3   **(See color insert following page 20.)** Dynamic lighting can be achieved with the use of multiple colored sources in the same system.

from fluorescent tubes in known products, up to 50% of the light is lost. Incandescent lamps with filters can also be used for dynamic lighting, but the ones that have dominated the field of dynamic lighting in recent years are LEDs (light-emitting diodes; both LEDs and incandescent light sources are easy to dim compared to discharge lamps).

Another popular application of systems for dynamic lighting is simulation of daylight in spaces that have no access to natural light (no windows). Dynamic lighting systems can reproduce the changes in color temperature that take place during the day, creating a feeling of time passing and of natural changes.

There have been various attempts to develop lighting systems that give the option of controlling the color temperature using only one light source. One idea is the use of two fluorescent powders/phosphors coated on the same tube and their selective excitation by changing the mode of operation of the lamp and, hence, the emission spectrum of the active medium.

The switching from steady state to pulse operation, for example, affects the relative intensities of mercury's emission lines because the mean electron energy is also affected. If different lines can excite different phosphors, then controlling the current means one can control the final emission spectrum.

Based on the same idea of switching the modes of operation, one can also use more than one active medium and control which one is excited, as different media have different energy level structures (or select between active medium and buffer gas). Controlling which atoms or molecules are excited by varying the mode of operation also means that one can control which phosphor is excited. In case a mixture of active media is employed, then by varying the current intensity one can control their diffusion through cataphoresis.

Regardless of the number or types of light sources, dynamic lighting systems offer flexibility and numerous options. Whether preprogrammed or set by the user, dynamic lighting systems can be used to accommodate many lighting needs and control the surrounding lighting at will.

CHAPTER 7

Luminaires

T HE LAMP IS THE heart of every lighting fixture, but a complete luminaire usually includes most of the following: reflectors for directing the light, optics for diffusing the light, an aperture (with or without a lens), the housing for lamp alignment and protection, ballast, if required, and connection to a power source. There is a great variety of luminaries, and the choice is made depending on both the lamp to be used and esthetic criteria. For incandescent lamps, with the exception of spot halogen lamps, all luminaires have a decorative design and character. Various types of reflectors are used in the design of luminaires, such as flat, elliptical, parabolic, spherical, and a combination of these (Figure 7.1).

For fluorescent lamps, the luminaires not only employ reflectors so that light is directed toward the desired work or living space, but they are also designed in such a way that the generated heat is dissipated and carried away. For both high- and low-pressure discharge lamps, the luminaires must also be able to accommodate the electronic gear (Figure 7.2).

When choosing a luminaire, one has to take into account its efficiency, which is defined as the ratio of the luminous flux of the luminaire over the luminous flux of the lamp(s). This is also known as the light output ratio (LOR). Another coefficient is the coefficient of utilization (CU), which is a measure of the efficiency of a luminaire in transferring luminous energy to the working plane in a particular area. A CU measures the light actually reaching the desired plane as a percentage of the total light produced by the fixture.

For the entertainment industry and displays, the lamps that dominate are those of high power and white light emission such as incandescent/halogen,

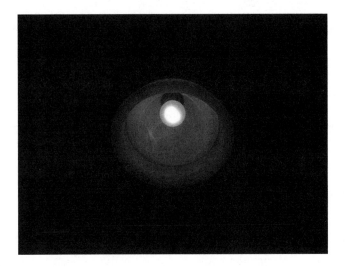

FIGURE 7.1 General-purpose luminaire.

xenon high pressure and, in some cases, carbon arc. For the emission of colored light, various filters are used in front of the light source, although with the ever-increasing use of LEDs, filters are no longer needed.

Luminaires can be broadly separated into two categories: *floodlights*, which illuminate a wide area (diffuse light) such as the PAR luminaires

FIGURE 7.2 Luminaire for linear fluorescent tubes.

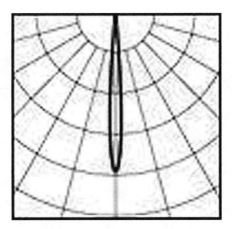

FIGURE 7.3   Characteristic beam profiles for spotlights.

(parabolic aluminized reflector), and *spotlights* (sometimes known as "Profiles"), which produce a narrower, more controllable light beam (focused light). The diagrams in Figures 7.3 and 7.4 show examples of the radial distribution of the emitted light in each case.

Floodlight luminaires are also categorized as circular, which best suit incandescent lamps (their diameter is proportional to the wattage of the light source used) and rectangular, which best suit high-pressure discharge lamps (see Figures 7.5 and 7.6).

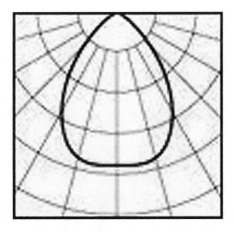

FIGURE 7.4   Characteristic beam profiles for (left) spotlights and (right) floodlights.

FIGURE 7.5   Floodlight luminaire for high-pressure discharge lamps.

Luminaires are also categorized depending on whether the luminaire is totally exposed or concealed behind a ceiling or wall. The ceiling-mounted version is often called a downlight.

Some examples of recessed fixtures include cans, which is a general term for inexpensive downlighting products recessed into the ceiling,

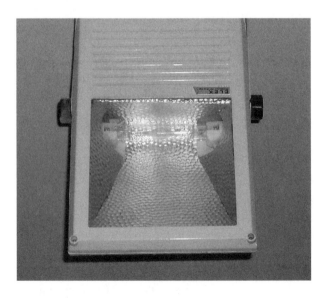

FIGURE 7.6   Floodlight luminaire for high-pressure discharge lamps.

FIGURE 7.7   Exposed/pendant lights.

and troffer luminaires, which refer to recessed fluorescent lights (the word comes from the combination of trough and coffer). Examples of exposed luminaires include the chandelier, the pendant lights suspended from a ceiling usually with a chain (Figure 7.7), the under-cabinet lights that are mounted below kitchen wall cabinets, high and low bay lights typically used for general lighting in industrial buildings, and strip lights, which are usually long lines of fluorescent lamps used in a warehouse or factory.

There is, of course, a whole range of luminaires used in outdoor lighting to illuminate streets, parking lots, building exteriors (Figure 7.8), and architectural details. These luminaires can be pole-mounted to illuminate streets and parking lots or mounted in the ground at low levels for illuminating walkways.

Some of the most expensive and sophisticated lighting systems are those used in professional applications such as stage lighting. Here again,

FIGURE 7.8 Luminaire used in outdoor lighting and specifically in building exteriors.

the categorization takes place depending on whether the luminaire is used for floodlighting or as a spotlight. Some widely used floodlight systems are the following.

The parabolic aluminized reflector lights, or PAR lights, or PAR cans, which are nonfocusable instruments: PAR cans consist mainly of a metal cylinder with a sealed-beam parabolic reflector lamp at one end. Like an old-fashioned automotive headlight, the reflector is integral to the lamp, and the beam spread of the unit is not adjustable except by changing the lamp. The instrument throws an unfocused beam, the shape of which depends on what type of lamp is in the instrument. Color frames can be used with most cans by using the clips present on the front of the instrument. PAR lights have low cost,

FIGURE 7.9   Parabolic aluminized reflector.

light weight, are easily maintainable, and have high durability. See
Figure 7.9 for an example.

Strip lights, also known as cyclorama, are long housings typically
containing multiple lamps arranged along the length of the
instrument and emitting light perpendicular to its length. Lamps
are often covered with individual filters of multiple colors (usually the primary colors) controlled by a separate electrical dim-
mer circuit. Varying the intensity of the different colors allows for
dynamic lighting.

The ellipsoidal reflector floodlights, better known as scoop lights, are
circular fixtures that do not have any lenses. They have an incandescent lamp mounted inside a large parabolic metal reflector that
directs the light out of the fixture. Since they do not have any lens
system, they are cheaper than other fixtures. However, the downside
of this is that the light cannot be focused at all (even PARs allow
more control than scoops). Scoops can accommodate filters for tailoring the emitted spectrum.

Some known spotlight categories used widely in the entertainment
business are the Fresnel spotlights and the ellipsoidal reflector
spotlights.

The Fresnel spotlights consist of a mirrored parabolic reflector, an
incandescent lamp at the focus, and a round Fresnel lens. The
distinctive lens has a "stepped" appearance instead of the "full" or

FIGURE 7.10   Ellipsoidal reflector spotlight (ERS).

"smooth" appearance of those used in other lanterns, and it was actually built this way so that lighthouses could throw light farther. It is this lens that lends the instrument both its name and its characteristic of having a soft focus. Adjustments allow the focusing of the beam from spot to flood. Provisions for color frames are generally present on the front of fresnels. The lamp and reflector remain as a fixed unit inside the housing, and are moved back and forth to focus the light.

The ellipsoidal reflector spotlight, ERS or Profile (United Kingdom), is a common nonautomated luminaire that usually consists of an incandescent lamp, an elliptical reflector, and convex lenses. The ERS (Figure 7.10) is a flexible instrument that allows the focus and shaping of the beam. Various colors and patterns can also be produced with the appropriate filters and shutters. Large versions are used as followspots to pick out specific people or objects on a stage.

Finally, the use of automatic *moving lights* (DMX is the protocol for electronic control of lighting), as seen in Figure 7.11, is becoming increasingly widespread. The direction of the light beam is controlled by moving the luminaire or just the reflectors and lenses that

FIGURE 7.11   Smart lighting with total electronic control.

can rotate and tilt. All functions can be computer and remote controlled. Such instruments provide features as color changing, pattern changing, tilting, strobing, etc. These are the most complicated instruments available, and involve a considerable number of technologies. There are many designs of automated lighting systems from as many companies.

CHAPTER 8

Lasers

A CLASS OF LIGHT SOURCES that is not used for general lighting but for a wide range of other applications due to their unique characteristics is the laser (light amplification by stimulated emission of radiation). Lasers were developed out of the technology of masers (microwave amplification by stimulated emission of radiation).

The emitted laser light is a spatially coherent (waves of same frequency and phase), narrow, low-divergence beam. The laser's beam of coherent light differentiates it from light sources such as lamps that emit *incoherent* light, of random phase varying with time and position. Although laser light is usually thought of as monochromatic, there are lasers that emit a broad spectrum of light (short pulses) or simultaneously at different wavelengths.

The invention of the laser can be dated to 1958 with the publication of the scientific paper "Infrared and Optical Masers" by Arthur L. Schawlow and Charles H. Townes, both working for Bell Labs. That paper launched a new scientific field and opened the door to a multibillion dollar industry. The work of Schawlow and Townes, however, can be traced back to the 1940s and early 1950s, and their interest in the field of microwave spectroscopy, which had emerged as a powerful tool for working out the characteristics of a wide variety of molecules (maser technology was the first to be developed). Other important figures in the development of this technology are Gordon Gould, who first used the term *laser* in his paper titled "The LASER, Light Amplification by Stimulated Emission of Radiation," and Alexander Prokhorov, who independently worked on this technology.

Lasers work by imparting energy to atoms or molecules, so that there are more in a high-energy ("excited") state than in some lower-energy state;

this is known as a *population inversion*. When this occurs, light waves passing through the material stimulate more radiation from the excited states than they lose by absorption due to atoms or molecules in the lower state. This "stimulated emission" was also the basis of masers.

Essentially, a laser consists of an amplifying or gain medium (this can be a solid, a liquid, or a gas) and a system to excite the amplifying medium (also called a pumping system). The medium is composed of atoms, molecules, ions, or electrons, whose energy levels are used to increase the power of a light wave during its propagation. What must be achieved is a population inversion in the energy levels of the species followed by stimulated emission.

The pumping system creates the conditions for light amplification by supplying the necessary energy that will lead to population inversion. There are different kinds of pumping systems: optical (flashlamps, other lasers, etc.), electrical (discharge through the gas, electric current in semiconductors), or even chemical (reactions that lead to products with population inversion).

Theodore Maiman invented the world's first laser, known as the *ruby laser*, in 1960. A ruby crystal is made up of aluminum oxide doped with chromium atoms. In a ruby laser, a ruby crystal is formed into a cylinder. A fully reflecting mirror is placed on one end and a partially reflecting mirror on the other. A high-intensity lamp around the ruby cylinder provides the energy in the form of white light, which triggers the laser action. The lamp flash excites electrons in the chromium atoms to a higher energy level. Upon returning to their normal state, the electrons emit their characteristic ruby-red light. Mirrors at each end reflect the photons back and forth, stimulating other excited chromium atoms to produce more red light, continuing this process of stimulated emission and amplification. The photons finally leave through the partially transparent mirror at one end. The ruby laser is still used, mainly as a light source for medical and cosmetic procedures, and also in high-speed photography and pulsed holography.

The output of a laser may be continuous (known as *CW* for *continuous wave*) or pulsed. In the continuous mode, the population inversion required for lasing is continually maintained by a continuous pump source. In the pulsed mode of operation, one can achieve higher power outputs per pulse for the same average energy usage. In order to produce laser pulses, several techniques are used, such as Q-switching, mode-locking, and gain-switching.

In a Q-switched laser, the population inversion is allowed to build up by making the cavity conditions unfavorable for lasing. Then, when the pump energy stored in the laser medium is at the desired level, the system is adjusted, releasing the pulse. This results in high peak powers as the average power of the laser is packed into a shorter time frame. The Q-switch may be a mechanical device such as a shutter, chopper wheel, or spinning mirror/prism placed inside the cavity, or it may be some form of modulator such as an acousto-optic device or an electro-optic device: a Pockels cell or Kerr cell. The reduction of losses (increase of Q) is triggered by an external event, typically an electrical signal, so the pulse repetition rate can therefore be externally controlled. In other cases, the Q-switch is a saturable absorber, a material whose transmission increases when the intensity of light exceeds some threshold. The material may be an ion-doped crystal such as Cr:YAG, which is used for Q-switching of Nd:YAG lasers, a bleachable dye, or a passive semiconductor device. Initially, the loss of the absorber is high, but still low enough to permit some lasing once a large amount of energy is stored in the gain medium.

Mode-locking is a technique by which a laser can be made to produce pulses of light of extremely short duration, on the order of picoseconds (10^{-12} s) or femtoseconds (10^{-15} s). The production of such short pulses has allowed exploration of really fast physical and chemical events (such as chemical reactions) as well as the study of nonlinear effects in optics. The basis of the technique is to induce a fixed phase relationship between the modes of the laser's resonant cavity. The laser is then said to be *phase-locked* or *mode-locked*. Interference between these modes causes the laser light to be produced as a train of pulses. A pulse of such short temporal length has a spectrum that contains a wide range of wavelengths. Because of this, the laser medium must have a broad enough gain profile to amplify them all. An example of a suitable material is titanium-doped, artificially grown sapphire (Ti:sapphire).

Yet another method of achieving pulsed laser operation is to pump the laser medium with a pulsed pumping source, which can be a flashlamp or another laser. Such a technique is common in the case of dye lasers, where the inverted population lifetime of a dye molecule was so short that a high-energy, fast pump was needed. Pulsed pumping is also required for lasers that disrupt the gain medium so much during the laser process that lasing has to cease for a short period. These lasers, such as the excimer laser and the copper vapor laser, can never be operated in CW mode.

The aforementioned techniques have led to ever-decreasing pulse durations and ever-increasing peak powers per pulse. In recent years, certain facilities around the world have produced extreme values.

For example, The National Ignition Facility (NIF) at the Lawrence Livermore National Laboratory in Livermore, California, recently achieved 150,000 J in a single 10 ns pulse. One application of such lasers is to direct this energy into a tiny pellet containing deuterium in an effort to induce nuclear fusion for the development of an abundant source of energy.

At the Max Planck Institute for Quantum Optics in Garching, Germany, a pulse of just under 1 fs (femtosecond), was produced (this is less than the time photons take to complete a single oscillation cycle). In order to generate such a short pulse, the final result must be a white burst containing all colors from the visible region well into the ultraviolet region of the spectrum. This is very different from typical lasers, which emit a single color, but such short pulses cannot be so monochromatic due to the uncertainty principle.

The Lawrence Livermore National Laboratory produced pulses of over 1 PW (petawatt; 10^{15} W) a few years ago. Because laser light can be focused to a very small spot, the focused energy density reached the equivalent of 30 billion joules in a volume of 1 cm^3, far larger than the energy density inside stars.

The Mid-Infrared Advanced Chemical Laser (MIRACL), at the High Energy Laser Systems Test Facility at White Sands Missile Range, New Mexico, achieved more than 1 MW (megawatt) of continuous output power. Because the power is so high, it is operated only for seconds at a time, producing several megajoules of energy in a single burst.

Lasers are categorized based on their safety level, which is represented by a class number. Classification for continuous lasers can appear like this:

- Class I is considered safe because usually the laser and light are enclosed (CD players).

- Class II is also considered safe because the blink reflex of the eye will prevent damage (lasers of up to 1 mW power).

- Class IIIa/3R lasers are usually up to 5 mW and involve a small risk of eye damage as staring into such a beam for several seconds is likely to cause minor eye damage.

- Class IIIb/3B can cause immediate, severe eye damage upon exposure (lasers of up to 500 mW, such as those in CD and DVD writers).

- Class IV/4 lasers are the most dangerous and can cause skin burns. Caution must be exercised even when dealing with scattered light. Most industrial and scientific lasers are in this class (Figures 8.1–8.3).

FIGURE 8.1 **(See color insert following page 20.)** Laser light emissions (Nd: YAG 532 nm) in research application.

Another method of categorization of lasers is based on the gain medium. Table 8.1 lists some of the best-known lasers along with some important information about them.

For most of the lasers listed, light is generated by methods already described, such as flow of electricity in semiconductors and electric discharges in a gas. Another type of laser is called the excimer laser, which is based on chemiluminescence. In excimer lasers, a chemical reaction takes place in which the product of the reaction is at a higher energy state than the reactants (population inversion).

FIGURE 8.2   Lighting effect with lasers.

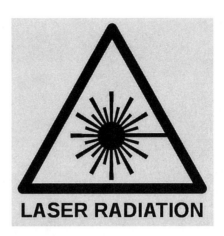

FIGURE 8.3 Standard sign warning of laser operation.

Noble gases such as krypton and xenon do not readily form chemical compounds but when their atoms are excited (as during an electrical discharge), pseudomolecules are formed of two noble gas atoms (dimers) or a noble gas atom and a halogen atom (complexes) such as fluorine or chlorine. The pseudomolecules under excitation (excited dimer = excimer or excited complex = exciplex) easily release the excess energy, returning to an unstable low energy level leading to their dissociation and release of free atoms again. These are the right conditions for population inversion, which is essential for the functioning of lasers. Depending on the type of pseudomolecules, the laser is called an excimer or an exciplex laser. In the case of Xe_2, the radiation emitted is at 172 and 175 nm (the same process and wavelengths we find in low-pressure xenon lamps), while other known wavelengths of laser emissions are 351 nm (XeF), 308 nm (XeCl), and 248 nm (KrF).

The same principle applies (without the flow of electricity) in the way light sticks work. These sticks, used in emergency situations or just for decoration, contain various chemical compounds isolated from each other. By breaking the stick, the different compounds mix, and the result of each chemical reaction is the emission of light. This method of creating light is very inefficient (about 1%), but in various enzymatic reactions taking place in living organisms such as fireflies, the efficiency can reach 90%.

The technology of lasers is applied in so many areas of human activity that it is impossible to imagine life without them. Laser light through optical fibers carries tremendous amounts of information (such as telephone conversations and computer connections). The telecommunication speeds

TABLE 8.1   Various Types of Lasers with Emissions in the Visible and their Applications

Gain Medium and Type of Laser	Wavelength (nm)	Pumping Method	Applications
Helium-neon/ gas	543, 632.8	Electrical discharge	Holography, spectroscopy, projections, shows, and displays
Argon ion/gas	450–530 (488 and 514.5 most intense)	Electrical discharge	Lithography, projections, shows, and displays
Krypton ion/gas	335–800	Electrical discharge	Scientific research, white lasers, projections, shows, and displays
Xenon ion/gas	Multiple emission lines in UV-Vis-IR	Electrical discharge	Scientific research
Dye lasers	300–1000	Laser, discharge lamp	Spectroscopy, skin treatment
HeCd laser metallic vapors	440, 325	Electrical discharge	Scientific research
Ruby/solid state	694.3	Discharge lamp	Holography, tattoo removal
Nd:YAG/solid state	1064, 532	Discharge lamp, diode laser	Material processing, medical procedures, scientific research, pumping other lasers, projections, shows, and displays
Ti:sapphire/ solid state	650–1100	Laser	Spectroscopy, LIDAR
Cr:Chrysoberyl (Alexandrite)/ solid state	700–820	Discharge lamp, diode laser	Skin treatment, LIDAR
Laser semiconductor diode	Multiple emission lines in Vis-IR	Electric current	Telecommunications, printing, holography, 780 nm AlGaAs for CD scanning, projections, shows, and displays

of today are largely due to laser technology. Supermarket checkout scanners, CDs, DVDs, laser holograms, and laser printers are just a few of the countless everyday technologies that rely on lasers. Industrial lasers cut, drill, and weld materials ranging from paper and cloth to diamonds and exotic alloys, far more efficiently and precisely than metal tools. Used in millions of medical procedures every year, lasers reduce the need for general anesthesia. The heat of the beam cauterizes tissue as it cuts, resulting in almost bloodless surgery and less infection. For example, detached retinas cause blindness in thousand of people each year. If caught early enough, a laser can "weld" the retina back in place before permanent damage results.

TABLE 8.2   The Efficiency of Lasers is Usually Too
Low, with the Exception of Carbon Dioxide Lasers,
Which are Used in the Industry for Material Processing

Laser	Efficiency (%)
Argon ion	0.001–0.01
Carbon dioxide	5.0–20.0
Excimer	1.5–2.0
GaAlAs semiconductor	1.0–10
Helium-Neon	0.01–0.1
Nd:YAG	0.1–1.0
Ruby	0.1–1.0

Before any other application, lasers were used for scientific research. At first, similar to masers, they were used to study atomic physics and chemistry. However, uses were soon found in many fields. For example, focused laser beams are used as "optical tweezers" to manipulate biological samples such as red blood cells and microorganisms. Lasers can cool and trap atoms, and create a strange new state of matter (the Bose–Einstein condensate) that probes the most fundamental physics. Over the long run, none of the uses of lasers is likely to be more important than their help in facilitating new discoveries, with unforeseeable uses of their own.

BIBLIOGRAPHY

Townes, C.H. (1999), *How the Laser Happened: Adventures of a Scientist*, Oxford University Press, New York, http://www.laserfest.org.

Silfvast, W.T. (1996), *Laser Fundamentals*, Cambridge University Press, United Kingdom.

Wilson, J. and Hawkes, J.F.B. (1987), *Lasers: Principles and Applications*, Prentice Hall International Series in Optoelectronics, Prentice Hall, New Jersey.

CHAPTER 9

Closing Remarks

THIS BOOK HAS GIVEN an overview of the three main technologies that gave birth to the numerous product families one finds in the market today. Electrical incandescence, electrical gas discharges, and semiconductor light-emitting diodes are the three dominant technologies that have lit up our world for more than a century. The spread of electricity and the development of power grids, which started with Edison, changed the way we illuminate our surroundings and created a new industry. In this book, we examined each technology separately and presented different families of light sources belonging to each category. In every case, there were many advantages and disadvantages to be discussed and different applications that a lamp could serve.

The attractive features of incandescent lamps such as the high color rendering index of the white light they emit and therefore the near-perfect color reproduction of the illuminated objects; or the robustness and long life of the light-emitting diodes; or the efficiency of electrical discharge lamps. Incandescent lamps and products based on LEDs still produce less than 30 lm/W, while products based on electrical discharges produce more than 40 lm/W, reaching a value of 200 in the case of low-pressure sodium discharges.

It is not a surprise that each technology has their own attractive characteristics and dominates different areas, because if a light source product or technology did not offer unique or competitive features, then it probably would have become obsolete, retired, and most likely not remembered in any detail except in a historical footnote next to gas and oil lamps.

An ideal artificial light source is difficult to develop and certainly does not exist today. For that reason, all or at least most technologies described in this book will continue to serve us unless a revolution in the light source research field takes place.

But what would an ideal light source be like? What would be its characteristics? Let us look at a list of features, and it will become clear that each technology examined has some but not all the features.

High luminous efficiency—Currently, white sources reach 120 lm/W, while the monochromatic low-pressure sodium lamp reaches 200 lm/W. A source that exceeds those limits is desirable and will have a high impact from an energetic and economic standpoint.

Long lifetime—Induction lamps (low-pressure mercury and high-pressure sulfur) exceed 20,000 h, while LEDs exceed 50,000 h. A technology that consistently exceeds 150,000 h is a realistic target.

Near-perfect color rendering—Incandescent lamps are assigned a value of 100, which several phosphors used with the other technologies can approach.

Color temperatures—Warm white is preferred by the majority of consumers, but the possibility of controlling the tones in a dynamic system would be ideal.

Avoidance of flickering and of electromagnetic interference with other electronic equipment

Attractive shape and form—Compact, light, strong, attractive, exchangeable with other types of lamps and fitting existing infrastructure (manufacturing and power supplies).

Environmentally friendly—The avoidance of toxic, harmful, or rare material (such as mercury in fluorescent lamps or indium in LEDs) as well as recyclable.

Useful emitter—The minimization of wasted energy in the form of heat or UV unless a special application requires it.

Low cost—Cheap for the consumer but profitable for the industry.

Looking at the foregoing criteria and desired features, it becomes apparent that the existing technologies satisfy several of them but no product can satisfy all of them; it will be quite a challenge to develop one that does. At the moment, an improvement in one characteristic might lead to compromise in another, so one can only use the products depending on one's needs and applications. The world and the industry would definitely change with such an ideal technology.

But what are the current developments? And what are the most pressing matters for researchers and developers?

Without a doubt mercury plays a central role in the market of light sources, as it is found in the majority of discharge lamps, from low-pressure mercury fluorescent lamps to high-pressure high-intensity discharges (HID) to super high-pressure projection lamps. Even in high-pressure sodium lamps and high-pressure metal halide lamps, mercury is usually added and acts as one of the main active media.

Light source scientists and engineers today are asked to not only push the upper limit of light source efficiency to higher values but also to eliminate mercury and search for a new material that will serve as an efficient radiator. The replacement of mercury in lighting products is a requirement that stems from long-lasting environmental concerns, particularly mercury's organic form, which is hazardous and life threatening. The replacement of mercury is a requirement of government and has found a new frame these days in the form of government official directives (the well-known RoHS directives, where the acronym stands for Reduction of Hazardous Substances). Apart from mercury, the use of lead, cadmium, and hexavalent chromium is also not allowed. The immediate need to replace mercury for environmental reasons will probably have some dramatic effects on the way low-pressure discharge lamps, such as the fluorescent lamp, are developed. The molecular low-pressure discharge lamps described in this book could prove to be a dynamic new forerunner.

The target for the efficiency and/or efficacy is, of course, for all scientists in the field, to reach the same value or higher than what has been achieved with the use of current products (for example, fluorescent lamps that can reach 120 lm/W, although there is usually a trade-off between efficacy and color rendering index depending on the phosphor chosen for different applications).

LED technology is the least researched as it was the latest one to penetrate the market. Most of the advances and areas of research in LED technology have been discussed in detail in Chapter 5. From the inner mechanisms of the crystals to the way they are packaged and manufactured, these are all areas where improvements will probably take place soon.

The near future of lighting will be defined by developments in solid-state lighting and the rapid increases in LED and OLED efficacies. Lamps based on the principle of discharges in high-pressure vapors already show high luminous efficacies and a wide range of color temperatures and rendering indices, so they will also play an important role in the decades ahead.

The technology of quantum dots, new advanced phosphors, electro-candoluminescence described earlier in this book, pulsing operation, and other methods for converting and manipulating spectra of existing light sources will offer new solutions if developed and might lead the whole scientific community to reexamine and rethink all existing technologies too. The efficiency of all existing technologies can benefit from such developments.

Many of the light sources mentioned are the result of combining different existing technologies. Starting with high-pressure mercury discharges, other products were developed, including metal halide lamps. The inductive operation principle of mercury fluorescent lamps was applied to sulfur lamps. In high-pressure xenon or sodium lamps, mercury has been added. The hybrid mercury lamp is a combination of the incandescent principle and mercury high-pressure discharge. The polycrystalline alumina burner (PCA) that was first developed for sodium lamps found use in the development of the metal halide family of lamps. All these examples show that although the field of research and development of light sources is waiting for a new revolution in methods for creating light, until that revolution takes place, the field will be dominated by combinations of existing technologies that will satisfy new needs or applications and the optimization or improvement of existing products.

Appendices

Appendix A

Tables of indicative or maximum values for various lamp characteristics and their applications

	Incandescent	Halogen
Efficacy (lm/W)	20	30
Power (W)	15–1,000	5–2,000
Color temperature (K)	2,800	3,000–3,500
Color rendering index	100	100
Lifetime (h)	1,000	2,000–5,000
Applications	Indoor reception spaces, shops, homes	Indoor reception spaces, shops, homes

	Fluorescent	CFL
Efficacy (lm/W)	up to 120	65
Power (W)	5–125	5–55
Color temperature (K)	Wide	Wide
Color rendering index	up to 99	85
Lifetime (h)	15,000–35,000	1,000
Applications	Public and outdoor spaces, offices, shops, industry, LCD screens	Public and outdoor spaces, offices, shops, homes, indoor reception spaces

	Sodium Low Pressure	Sodium High Pressure
Efficacy (lm/W)	200	50–150
Power (W)	35–180	35–1,000
Color temperature (K)	1,700	2,000–3,000
Color rendering index	0	20–85
Lifetime (h)	20,000	10,000–30,000
Applications	Public and outdoor spaces, security lights, street lighting	Public and outdoor spaces, security lights, street lighting

	MH	MH (PCA)
Efficacy (lm/W)	90	100
Power (W)	70–2,000	35–400
Color temperature (K)	Wide	Wide
Color rendering index	70–90	up to 90
Lifetime (h)	10,000	10,000–20,000
Applications	Public and outdoor spaces, offices, shops, industry, high levels of luminous flux, large indoor spaces, warehouses	Public and outdoor spaces, offices, shops, industry, high levels of luminous flux, large indoor spaces, warehouses

	Hg High Pressure	Hg Very High Pressure
Efficacy (lm/W)	60	60
Power (W)	50–1000	100–250
Color temperature (K)	3,000–4,000	7,500
Color rendering index	15–55	up to 60
Lifetime (h)	10,000–30,000	10,000
Applications	Public and outdoor spaces, high levels of luminous flux, large indoor spaces, industry, warehouses	Projectors, daylight simulation

	LED	Hybrid
Efficacy (lm/W)	130	20
Power (W)	0.1–7	100–500
Color temperature (K)	Wide range	3,300–3,700
Color rendering index	up to 90	50–70
Lifetime (h)	100,000	10,000
Applications	Signs, remote controls, fiber-optic communication, remote controls displays, decoration, advertisement	Public and outdoor spaces

	Hg Inductive	Sulfur
Efficacy (lm/W)	80	95
Power (W)	55–165	1,000–6,000
Color temperature (K)	2,700–4,000	6,000
Color rendering index	80	80
Lifetime (h)	60,000–100,000	20,000
Applications	Inaccessible spots due to long life	Inaccessible spots due to long life, high levels of luminous flux

	Xe Excimer	Xe High Pressure
Efficacy (lm/W)	30	>30
Power (W)	20–130	1–15 kW
Color temperature (K)	8,000	>6,000
Color rendering index	85	>90
Lifetime (h)	100,000	2,000
Applications	Photography, laser construction	Projectors, daylight simulation

Lux (Illuminance)	Kind of Environment or Activity
20–50	Outdoor workspace
30–150	Short stay rooms
100–200	Workrooms of noncontinuous usage
200–500	Work with simple visual requirements
300–750	Work with average visual requirements
500–1,000	Work with high visual requirements
750–1,500	Work with very high visual requirements
1,000–2,000	Work with special visual requirements
>2,000	Extremely visually accurate work

Lux (Illuminance)	Kind of Outdoor Space
1–10	General storage spaces, security
10–50	Car parking space, cargo transfers
50–500	Sales, sport events, advertisement
500–1,000	Sport events with audience
1,000–2,000	Sport event with television coverage

Color Rendering Index (R_a)	Kind of Environment or Activity
>90	Galleries, printing facilities
80–90	Homes, restaurants, fabric industry, museums
60–80	Office, schools, light industry
40–60	Heavy industry, corridors, stairs
20–40	Outdoors

Conservation Category	Maximum Surface Illuminance (lux)	Maximum Annual Exposure (lux-hours)
Objects without photosensitivity (metals, jewels, glass)	No limit	No limit
Average photosensitivity (wood, paintings, bones, skin)	200	600,000
High photosensitivity (fabrics, prints, drawings, watercolorings, botanical samples)	50	150,000

Appendix B: The Different Sections of the Electromagnetic Spectrum

Red, orange, yellow, green, and blue each exhibit unique frequencies and, consequently, wavelengths. While we can perceive these electromagnetic waves in their corresponding colors, we cannot see the rest of the electromagnetic spectrum.

Most of the electromagnetic spectrum is invisible, and exhibits frequencies that traverse its entire breadth. Exhibiting the highest frequencies are gamma rays, x-rays, and ultraviolet light. Infrared radiation, microwaves, and radio waves occupy the lower frequencies of the spectrum. Visible light falls within a very narrow range in between.

Radio waves	10^4–10^{-2} m/10^4–10^{10} Hz	
	Ultra-low frequency (ULF)	3–30 Hz
	Extremely low frequency (ELF)	30–300 Hz
	Voice frequencies (VF)	300 Hz–3 kHz
	Very low frequency (VLF)	3–30 kHz
	Low frequency (LF)	30–300 kHz
	Medium frequency (MF)	300 kHz–3 MHz
	High frequency (HF)	3–30 MHz
	Very high frequency (VHF)	30–300 MHz
	Ultra high frequency (UHF)	300 MHz–3 GHz
	Super high frequency (SHF)	3–30 GHz
	Extremely high frequency (EHF)	30–300 GHz
	Shortwave	MF, HF

	Television	VHF, UHF
	Microwaves	30 cm–1 mm /1–300 GHz
Infrared	10^{-3}–10^{-6} m/10^{11}–10^{14} Hz	
	Far	1000–30 μm
	Mid	30–3 μm
	Near	3–0.75 μm
Visible	5×10^{-7} m/2×10^{14} Hz	
	Red	770–622 nm
	Orange	622–597 nm
	Yellow	597–577 nm
	Green	577–492 nm
	Blue	492–455 nm
	Violet	455–390 nm
Ultraviolet	10^{-7}–10^{-8} m/10^{15}–10^{16} Hz	
	UV-A (less harmful)	400–315 nm
	UV-B (harmful, absorbed by ozone)	315–280 nm
	UV-C (more harmful, absorbed by air	280–100 nm
	Near UV ("black light")	400–300 nm
	Far UV	300–200 nm
	Vacuum UV	200–100 nm
X-rays	10^{-9}–10^{-11} m/10^{17}–10^{19} Hz	
Gamma rays	10^{-11}–10^{-13} m/10^{19}–10^{21} Hz	

Appendix C

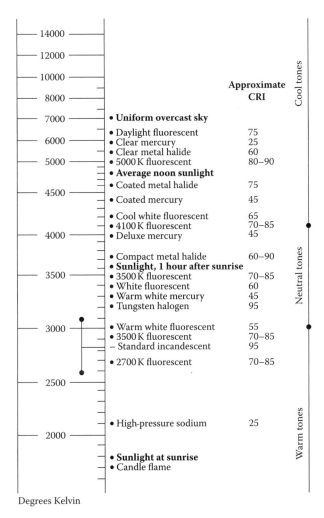

FIGURE C.1 Color temperatures and color rendering indices of various light sources.

Appendix D: List of Phosphor Powders for Fluorescent Lamps

- $(Ba,Eu)Mg_2Al_{16}O_{27}$, blue phosphor for trichromatic fluorescent lamps

- $(Ce,Tb)MgAl_{11}O_{19}$, green phosphor for trichromatic fluorescent lamps

- $(Y,Eu)_2O_3$, red phosphor for trichromatic fluorescent lamps

- $(Sr,Eu,Ba,Ca)_5(PO_4)_3Cl$, blue phosphor for trichromatic fluorescent lamps

- $(La,Ce,Tb)PO_4$, green phosphor for trichromatic fluorescent lamps

- $Y_2O_3{:}Eu$, red phosphor (611 nm), for trichromatic fluorescent lamps

- $LaPO_4{:}Ce,Tb$, green phosphor (544 nm) for trichromatic fluorescent lamps

- $(Sr,Ca,Ba)_{10}(PO_4)_6Cl_2{:}Eu$, blue phosphor (453 nm) for trichromatic fluorescent lamps

- $BaMgAl_{10}O_{17}{:}Eu,Mn$, blue-green (456/514 nm)

- $(La,Ce,Tb)PO_4{:}Ce,Tb$, green (546 nm), phosphor for trichromatic fluorescent lamps

- $Zn_2SiO_4{:}Mn$, green (528 nm)

- $Zn_2SiO_4{:}Mn,Sb_2O_3$, green (528 nm)

- $Ce_{0.67}Tb_{0.33}MgAl_{11}O_{19}{:}Ce,Tb$, green (543 nm), phosphor for trichromatic fluorescent lamps

- Y_2O_3:Eu(III), red (611 nm) phosphor, for trichromatic fluorescent lamps

- $Mg_4(F)GeO_6$:Mn, red (658 nm)

- $Mg_4(F)(Ge,Sn)O_6$:Mn, red (658 nm)

- $MgWO_4$, soft blue (473 nm), wide range, deluxe

- $CaWO_4$, blue (417 nm)

- $CaWO_4$:Pb, blue (433 nm), wide range

- $(Ba,Ti)_2P_2O_7$:Ti, blue-green (494 nm), wide range, deluxe

- $Sr_2P_2O_7$:Sn, blue (460 nm), wide range, deluxe

- $Ca_5F(PO_4)_3$:Sb, blue (482 nm), wide range

- $Sr_5F(PO_4)_3$:Sb,Mn, blue-green (509 nm), wide range

- $BaMgAl_{10}O_{17}$:Eu,Mn, blue (450 nm), phosphor for trichromatic fluorescent lamps

- $BaMg_2Al_{16}O_{27}$:Eu(II), blue (452 nm)

- $BaMg_2Al_{16}O_{27}$:Eu(II),Mn(II), blue (450+515 nm)

- $Sr_5Cl(PO_4)_3$:Eu(II), blue (447 nm)

- $Sr_6P_5BO_{20}$:Eu, blue-green (480 nm)

- $(Ca,Zn,Mg)_3(PO_4)_2$:Sn, orange-pink (610 nm), wide range
  $(Sr,Mg)_3(PO_4)_2$:Sn, orange-pink white (626 nm), wide range, deluxe

- $CaSiO_3$:Pb,Mn, orange-pink (615 nm)

- $Ca_5F(PO_4)_3$:Sb,Mn, yellow

- $Ca_5(F,Cl)(PO_4)_3$:Sb,Mn, daylight

- $(Ca,Sr,Ba)_3(PO_4)_2Cl_2$:Eu, blue (452 nm)

- $3\ Sr_3(PO_4)_2.SrF_2$:Sb,Mn, blue (502 nm)

- $Y(P,V)O_4$:Eu, orange-red (619 nm)

- $(Zn,Sr)_3(PO_4)_2$:Mn, orange-red (625 nm)

- Y_2O_2S:Eu, red (626 nm)

- **(Sr,Mg)$_3$(PO$_4$)$_2$:Sn(II)**, orange-red (630 nm)

- **3.5 MgO . 0.5 MgF$_2$. GeO$_2$:Mn**, red (655 nm)

- **Mg$_5$As$_2$O$_{11}$:Mn**, red (660 nm)

- **Ca$_3$(PO$_4$)$_2$.CaF$_2$:Ce,Mn**, yellow (568 nm)

- **SrAl$_2$O$_7$:Pb**, ultraviolet (313 nm)

- **BaSi$_2$O$_5$:Pb**, ultraviolet (355 nm)

- **SrFB$_2$O$_3$:Eu(II)**, ultraviolet (366 nm)

- **SrB$_4$O$_7$:Eu**, ultraviolet (368 nm)

- **MgGa$_2$O$_4$:Mn(II)**, blue-green, black light

- **(Ce,Tb)MgAl$_{11}$O$_{19}$**, green

Appendix E

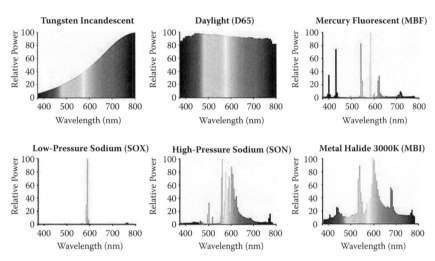

FIGURE E.1   Emission spectra of various lamps.

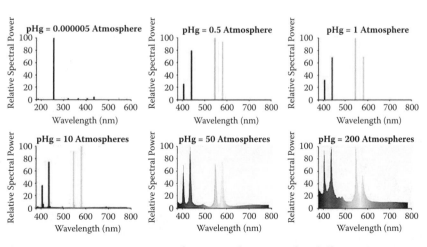

FIGURE E.2   Emission spectra of mercury lamps under different pressures.

Appendix F

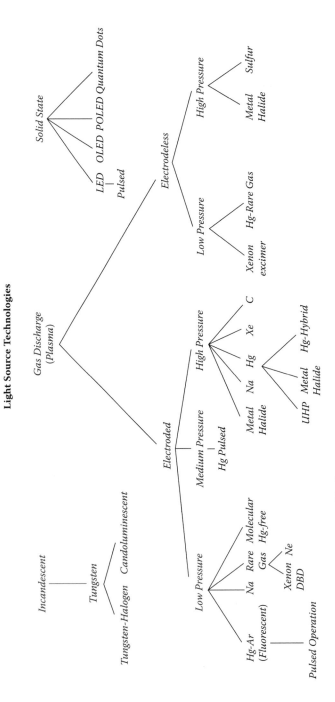

FIGURE F.1 Technology tree of light sources.

Index